Ed Noll's
Solid-State Circuit Files

Volume 1

Bipolar Transistor, FET, and Linear IC Circuits

by

Edward M. Noll

Howard W. Sams & Co., Inc.
4300 WEST 62ND ST. INDIANAPOLIS, INDIANA 46268 USA

FIRST EDITION
FIRST PRINTING—1980

International Standard Book Number: 0-672-21753-8
Library of Congress Catalog Card Number: 80-52231

Printed in the United States of America.

Preface

This book is the first volume of a two-volume set that consists of five separate circuit categories or "files" from which you can gain knowledge and practical experience. This volume covers bipolar and field-effect transistors and linear ICs. Each of the circuits designed has been built and checked out carefully by the author. The component values are given on each schematic diagram. An explanation and often a suggested experimental procedure accompany each circuit.

The new technician can begin with File 1-1 in Vol. 1 and proceed circuit after circuit to File 5-31 in Vol. 2, learning and gaining experience in solid-state electronics in a sensible and orderly manner. The material is arranged for progressive study as in a course with experiments.

If you wish to consider just one of the three major topics in this volume, you will find that the file itself is arranged for sequential study. If you are a teacher or experienced technician, you may wish to go directly to a specific circuit. In this case refer to the Table of Contents or Index.

All of the circuits in this volume have been built on solderless breadboards. This removes the tedium from building and experimenting with circuits. The circuits can be assembled quickly and made ready to go. The experiments in the book were powered with 6- and 12-volt lantern batteries.

Solderless circuit boards are readily available. An example is the Global Specialties Corporation Experimenter 300 board shown in Fig. 1. Two of these boards latched together permit the construction of the most elaborate circuits given in the book. The Experimenter 300 consists of five connected tie points per terminal. There are two separate rows of these tie-point terminals. The total number of terminals is 94. Down each side of the board is a single continuous strip of tie points to permit connection to the positive and negative supply voltage.

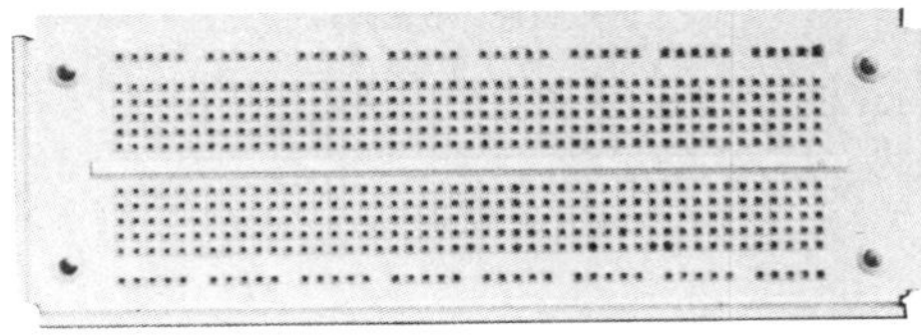

Courtesy Global Specialties Corp.

Fig. 1. Solderless circuit board.

Number 24 AWG hookup wire is used to make interconnections. Keep in mind that for the main section of the board each five points are tied together and for the bus strip there is a continuous connection from one end to the other. Integrated circuits are inserted across the two separate rows of tie-point terminals.

Every attempt was made to use commonly available parts, which can be found at ham radio and electronics parts stores. Good sources of integrated-circuit devices are the mail-order houses that advertise in the ham radio magazines and popular electronics periodicals. While specific part numbers are listed for many of the components, any equivalent by another manufacturer should work equally well.

May you learn more about solid-state electronics in a relaxing and enjoyable manner.

EDWARD M. NOLL

Contents

FILE 1

FILE 2

FILE 3

FILE 1

BIPOLAR TRANSISTOR CIRCUITS

BIPOLAR COMMON-EMITTER AMPLIFIER

A bipolar transistor is a three-element device consisting of emitter, base, and collector (Fig. 1-1). The emitter and collector elements are positioned on the two sides of the base. In this position the base is able to control the motion of charges between emitter and collector. The bipolar transistor has two junctions. The emitter junction, between the emitter and the base, is forward biased, and the collector junction, between base and collector, is reverse biased.

When the emitter junction is forward biased there is a motion of charges (current) across the junction. These charge carriers diffuse through the collector junction even though it is reverse biased. The charge carriers are attracted across the junction by the potential of the collector, and a substantial current results. As the base bias current is increased with more emitter-junction forward bias, the collector current rises. If the base current is made to vary with a signal, there results a similar and amplified change in the collector current. In the case of the common-emitter circuit of Fig. 1-1, there is a substantial voltage change across the output. This is greater than the voltage change applied to the input of the circuit. Therefore the common-emitter stage has voltage and current gains. The input and output voltages are 180° related.

The emitter junction of a bipolar transistor is forward biased. This junction has a high conductance and a significant base current. As a result the input resistance of a bipolar circuit is low, a factor that must be taken into consideration in terms of the loading influence it has on the preceding stage. The interstage coupling system must be designed to make a suitable match to the low input resistance of a bipolar stage. Likewise, the preceding stage must supply a signal of adequate power level if the stage is to operate in a normal manner.

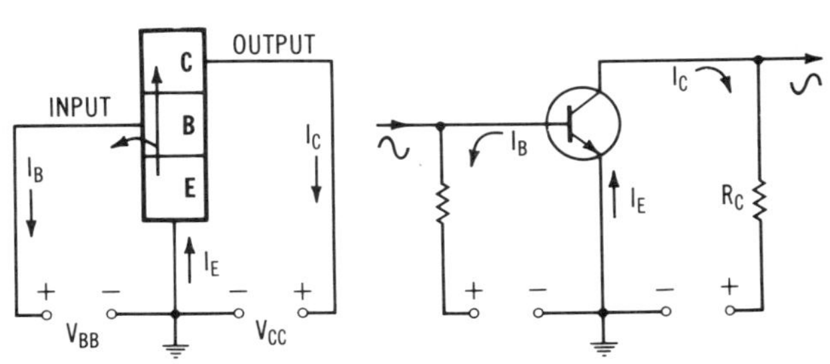

(A) Structure of transistor. (B) Circuit symbol of transistor.

Fig. 1-1. Basic bipolar transistor operation.

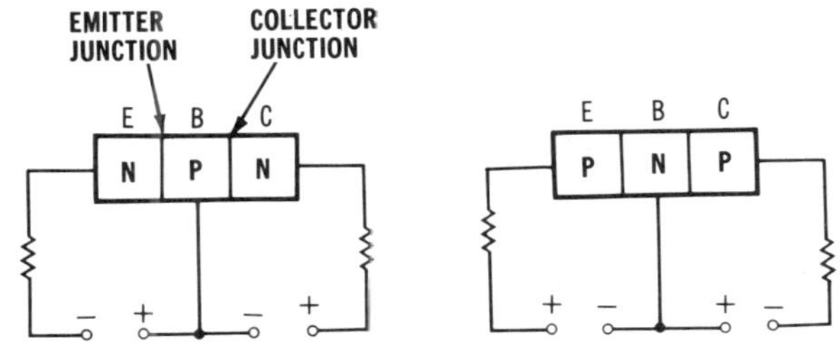

(A) Npn transistor biasing. (B) Pnp transistor biasing.

Fig. 1-2. Biasing of bipolar transistors.

There are two fundamental types of bipolar transistors: those with a p-type base and those with an n-type base (Fig. 1-2). For proper operation of an npn transistor a positive voltage is applied to the collector. As a result the collector junction is reverse biased, the base being negative with respect to the collector. The emitter junction must be forward biased. As a result the emitter must be made negative relative to the base. A positive bias on the base thus forward biases the emitter junction. Since the base is made positive with respect to the n-type emitter, the electrons of the emitter cross the junction into the base. These electrons diffuse through the base and are attracted by the positive voltage of the collector. Therefore they cross the collector junction and flow into the output circuit.

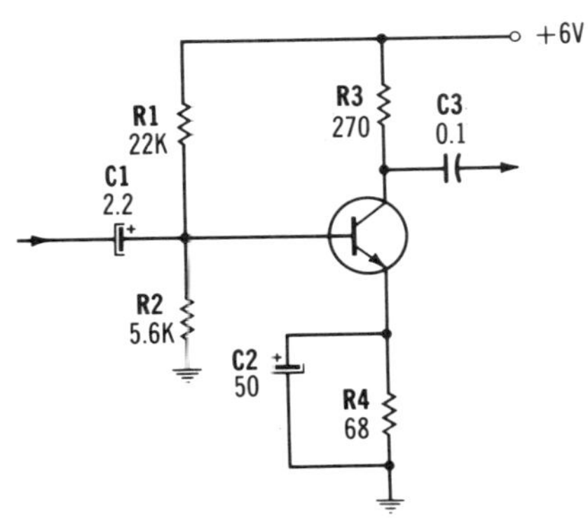

ALL RESISTORS 1/4W

Fig. 1-3. Common-emitter stage using general-purpose bipolar transistor.

Conversely, the pnp transistor must be arranged in a manner such that a negative voltage is applied to the collector. In so doing the collector junction is reverse biased. Likewise, the base must be made negative relative to the emitter in order to forward bias the emitter junction. In this case the charge carriers in the p-type emitter are positive charges, or holes. They are attracted by the negative biasing of the base and cross the emitter junction into the base. The positive charges diffuse through the base and are attracted by the negative potential of the collector. They cross the collector junction into the output circuit.

The practical circuit for a common-emitter amplifier, using a general-purpose bipolar transistor, is given in Fig. 1-3. Resistor *R3* is the collector load resistor and its value is selected to maintain a dc collector current that is approximately 1 milliampere. Resistor *R4* is an emitter resistor that provides dc stabilization. Capacitor *C2* in conjunction with resistor *R4* has a time constant that is long enough to minimize degeneration at low audio frequencies. Capacitors *C1* and *C3* also influence the low-frequency response. Their values are chosen to maintain a good response down to at least 100 hertz. The high-frequency response of the circuit is well over 20,000 hertz.

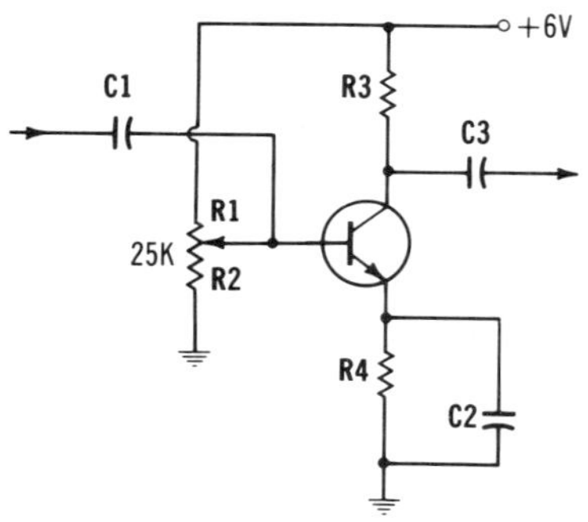

Fig. 1-4. Use of potentiometer to set values for *R1* and *R2*.

Resistors *R1* and *R2* set the operating base-bias current for the transistor. The values given for them set the bias correctly for the 2N2222 and Radio Shack 2009 general-purpose transistors. Other general-purpose transistors can be used with appropriate adjustment of the *R1* and *R2* values. One way to do this is to insert a potentiometer in the circuit as shown in Fig. 1-4. Connect an oscilloscope across the output. Apply a 1000-hertz audio signal to the input. Adjust the potentiometer to obtain a symmetrical and nondistorted audio output. Make certain that the input is not overdriven by the audio source. It may be necessary to insert a series dropping resistor between the audio source and the input side of capacitor *C1* to obtain a test signal of a level low enough.

If an oscilloscope is not available, a good approximation of proper operating conditions can be obtained by adjusting the potentiometer until the collector source voltage is about 3 volts. Note that this is one-half of the supply-voltage value. Remove power and measure the *R1* and *R2* sides of the potentiometer with an ohmmeter to obtain values of *R1* and *R2* for use in the schematic of Fig. 1-3.

COMMON-BASE AMPLIFIER

In the common-base configuration (Fig. 1-5) a signal is applied between the emitter and common and removed between the collector and common. The File 1-1 circuit was that of a common-emitter stage with the signal being applied to the base and removed at the collector. The common-base amplifier is capable of voltage gain but has a current gain that is less than unity. The common-emitter stage is capable of both voltage and current gain.

The common-base stage has a very low input impedance and a significantly higher output impedance. Therefore it can be used in any circuit where there must be a transformation between low input and high output impedance at the same time there is a possible voltage gain.

The schematic diagram of a practical common-base stage is given in Fig. 1-6. The same Radio Shack 2009 (similar to a 2N2222) transistor is used. The components are approximately the same but have been rearranged. Resistor *R1* now serves as the input resistance, being connected between emitter and common. Capacitor *C2* now appears across the base resistor, *R2*, with the combination of resistors *R2* and *R3* developing the necessary base bias current. It is necessary to increase the value of capacitor *C1* to 220 μF because of the much lower input impedance of the common-base stage. This step maintains the frequency response to a low limit of at least 120 hertz. The high-frequency response is well over 20,000 hertz.

In the common-base configuration the input and output voltages are in phase. In the previous common-emitter circuit the input and output voltages were 180° related.

In testing the circuit it is usually necessary to apply an input voltage higher than in the common-emitter circuit because of the loading effect of the low-impedance emitter input circuit. The voltage gain is approximately the same as that obtained with the common-emitter stage. Maximum undistorted output can be obtained by proper regulation of the values of resistors *R2* and *R3*. Again, a potentiometer can be used if you desire to obtain precise values and the highest undistorted output.

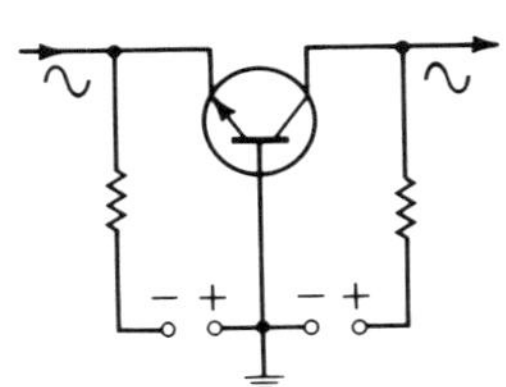

Fig. 1-5. Basic common-base configuration.

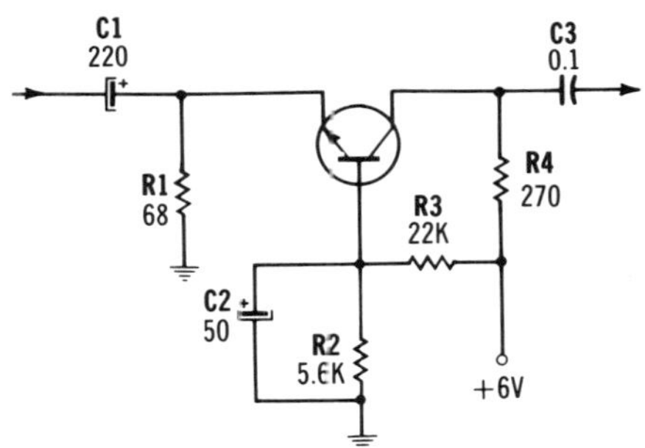

ALL RESISTORS ¼W

Fig. 1-6. Common-base circuit.

FILE 1-3

COMMON-COLLECTOR (EMITTER-FOLLOWER) STAGE

In the common-collector configuration the input signal is applied to the base, and the output is taken at the emitter (Fig. 1-7). The common-collector or emitter-follower stage is an effective impedance transformer. The input impedance is moderately high, the output impedance very low. Thus the stage is capable of current gain but not voltage gain. In a properly designed emitter-follower stage the output voltage is just slightly less than the input voltage. Such a stage is used mainly in matching a high-impedance signal source to a low-impedance load.

The output voltage is in phase with the input voltage. Another advantage of the common-collector stage is its capability of handling a high input-signal level without distortion while, at the same time, it places a minimum load on the source of the signal.

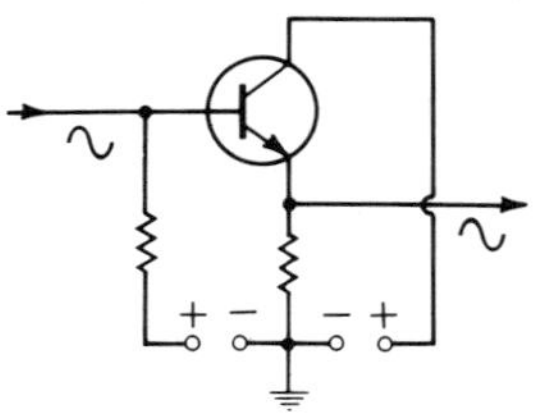

Fig. 1-7. Basic common-collector configuration.

The schematic of an emitter-follower stage is given in Fig. 1-8. The practical stage uses the Radio Shack 2009 (similar to the 2N2222) transistor. The components are approximately the same as used in the File 1-1 and File 1-2 schematics except that they are rearranged. The 270-ohm load resistor is now placed in the emitter circuit. The 50-μF capacitor is connected between the collector and ground to establish the common-collector configuration.

The stage will operate over a considerable range of audio input signal levels and has a low-frequency response that extends down to at least 100 hertz. The high-frequency response extends well over 20,000 hertz.

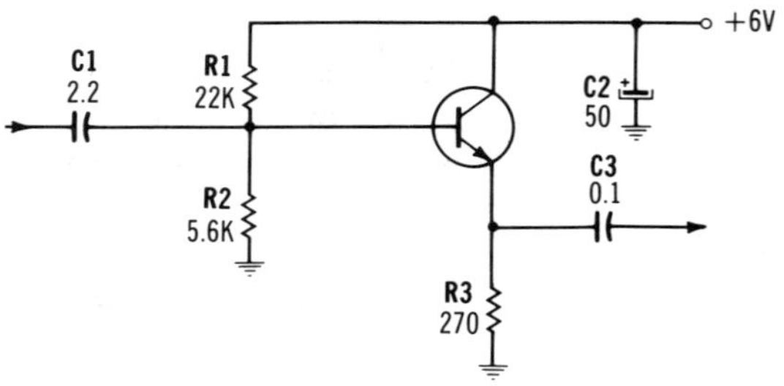

ALL RESISTORS ¼W

Fig. 1-8. Emitter-follower circuit.

TWIN-T AUDIO OSCILLATOR

A transistor can be made to operate as an oscillator if a signal from the output circuit is fed back in phase to the input. Since the common-emitter configuration already has an inherent 180° phase shift it is only necessary for a filter network to introduce an additional 180° shift. Note in Fig. 1-9 that the twin-T network is connected between the collector output and the base input of the transistor oscillator. Inasmuch as in-phase feedback occurs at only one particular frequency the frequency of oscillation depends on the constants of the twin-T network. The values shown in Fig. 1-9 produce an output frequency of approximately 1000 hertz.

The oscillator uses a 2N3904 transistor followed by a similar transistor connected as an emitter-follower to obtain a low-impedance output. Any reasonable load placed on the output will not change the frequency of oscillation.

The frequency of oscillation can be changed by substituting different values for capacitors *C1* and *C2*. A frequency of approximately 10,000 hertz can be obtained by using 500-pF capacitors; a 100-hertz output, by using 0.05-μF capacitors. A potentiometer inserted in place of resistor *R1* can be used to fine-tune the network to a specific frequency. Then a fixed value of resistor can be substituted. The output level can be changed over a limited range by adjusting the ohmic value of resistor *R6* in the emitter circuit of the oscillator.

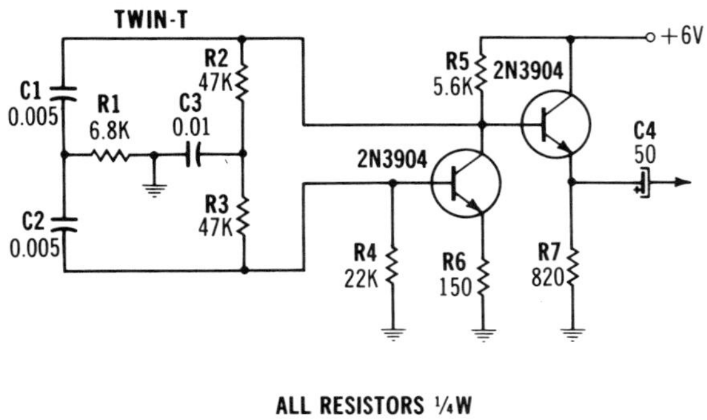

Fig. 1-9. Transistor audio oscillator.

FILE 1-5

SMALL-SIGNAL AUDIO AMPLIFIER

A stable small-signal, high-gain amplifier is shown in Fig. 1-10. Note that the collector of the input stage is direct coupled to the base of the output stage. Signal as well as current bias for the output stage is conveyed over this path. No interstage coupling capacitor is required. Stabilized bias for the input stage is obtained from the emitter circuit of the output stage by way of resistor *R2*.

The amplifier gain is excellent and a 3- to 4-volt peak signal can be developed across the output from an extremely low level input signal, such as that obtained from the output of a radio detector, microphone, etc. If an audio generator is used as a signal source to check out the amplifier, it will generally be necessary to use a very high value resistor in series with the input to reduce the signal to a sufficiently low level. The writer found it necessary to use a 2.5-megohm resistor in series with the input to reduce the output of his audio generator to a very weak signal.

The component values shown in Fig. 1-10 give good operation over voice-frequency range. Capacitor *C3* has a decided influence on the high-frequency response, and it can be reduced in value if an extended high-frequency performance is desired. Input capacitor *C1* and the emitter-filter capacitor, *C5,* influence the low-frequency response. Low-frequency response can be extended by increasing the value of either capacitor. A 220-μF capacitor for *C5* extends the response below 100 hertz.

Transistors that perform well in the circuit are the 2N3904 and 2N2222 types. Other general-purpose types can be used. Resistors *R1, R2,* and *R3* have major influence on the biasing of the two direct-coupled stages, and may require adjustment in value for other transistor types.

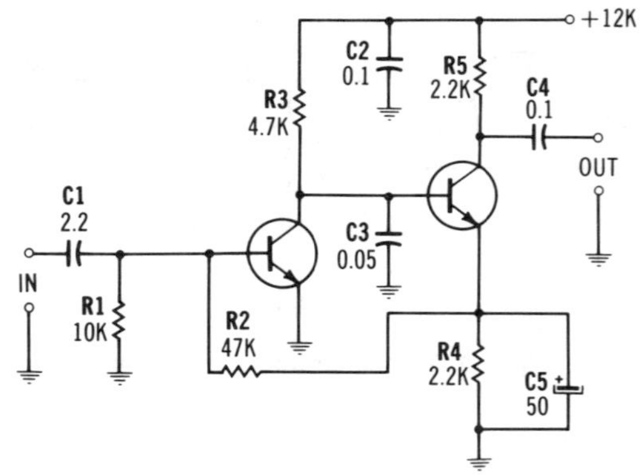

Fig. 1-10. Small-signal high-gain amplifier.

DARLINGTON CIRCUIT

The Darlington is a popular bipolar transistor circuit. It consists of two or more emitter-follower (common-collector) stages connected in cascade. The emitter follower has a high input impedance and a low output impedance. This impedance differential can be increased by cascading emitter-follower stages. Each stage provides additional current gain; consequently a very weak input current is able to produce a high output current. The overall voltage gain is less than unity but this is not serious in a well-planned circuit.

The circuit of Fig. 1-11 consists of a low-level, general-purpose input transistor followed by a higher-current output transistor. Note that the emitter of the input transistor is direct coupled to the base of the output transistor. The output transistor, too, is an emitter-follower stage. Both collectors are at ac ground potential.

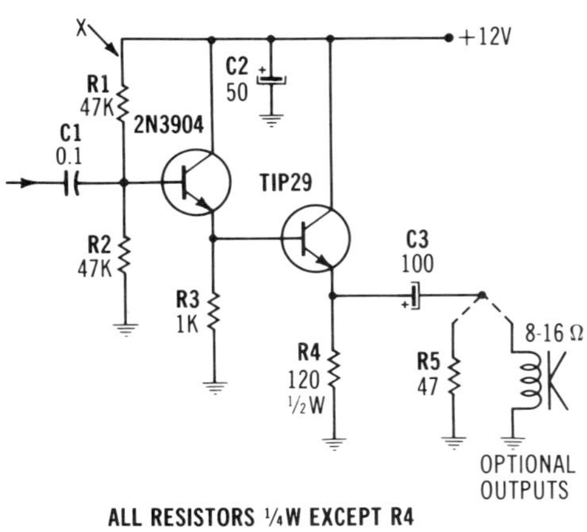

Fig. 1-11. Darlington circuit.

The input signal can be supplied from a relatively high impedance source, such as an audio generator or a common-emitter amplifier stage. A low-impedance output signal at about the same voltage level is developed at the emitter of the output transistor. In the author's test setup two types of load were alternately placed across the output: a 47-ohm resistor and a small 8-ohm speaker. There was very little reduction in voltage with the 47-ohm load, indicating the low output impedance of the Darlington circuit. Good speaker level was obtained even though there was a significant drop in the output voltage because of the very low speaker impedance.

The dc current gain was significant. An input bias current of a fractional part of a milliampere resulted in a dc emitter current of 35 to 40 milliamperes. The Darlington circuit is used widely in power control systems because of its ability to use a very minute input current to switch a high output current. You can check this out by connecting a voltmeter across the 120-ohm output resistor. Measure the voltage across *R4* and calculate the dc emitter current using Ohm's law. Now break the circuit at point *X* in the base circuit of the input transistor. This will remove the bias current. Note that the output reading drops to zero. This step proves that a very low input bias current results in a high dc output current.

AUDIO POWER OUTPUT STAGE

The audio power output stage of Fig. 1-12 consists of an emitter-follower input stage which drives a Class-A common-emitter output transistor. A Radio Shack 276-2018 can be substituted for the TIP-29 output transistor.

A miniature audio transformer matches the collector circuit of the output transistor from an 8-ohm speaker. Use a transformer (Radio Shack 273-1380) with a primary winding impedance of 1000 ohms and a secondary winding impedance of 8 ohms. If the transformer is center tapped, use the outer leads of the primary.

The emitter-follower input stage permits the source of signal to be of rather high impedance. The amplifier has good gain and it will be necessary to use a series dropping resistor when applying a test signal from an audio generator. If desired, a volume control can be connected across the input. It can be a 50-kilohm to 100-kilohm potentiometer with the left side of the input capacitor *C1* connected to the arm.

If low-gain operation is desired, capacitor *C4* in the emitter circuit of the output stage can be eliminated. Otherwise capacitor *C4* can be used to regulate the low-frequency response of the amplifier. The value shown provides good low-frequency response down to several hundred hertz. Capacitor *C3* across the primary winding of the output transformer can be used to regulate the high-frequency response. The value shown gives good response in excess of 10,000 hertz.

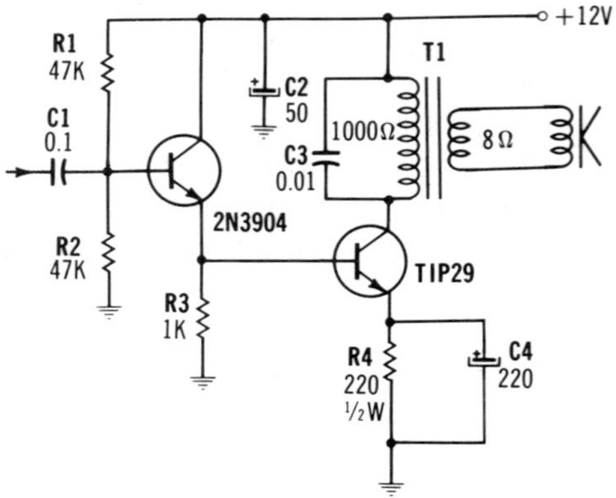

ALL RESISTORS ¼W EXCEPT R4

Fig. 1-12. Audio power output stage.

COMPLEMENTARY-SYMMETRY AMPLIFIERS

In a complementary-symmetry amplifier, npn and pnp transistors of similar characteristics are connected in series. The output is removed via a capacitor connected to their junction. An advantage of the complementary-symmetry connection is that a push-pull output is obtained without the necessity for a phase-inverter input. One transistor conducts on one alternation; the other transistor on the other alternation. Some power must be supplied by the driver, which must have the required power dissipation capability.

A simple driving stage is used in the circuit of Fig. 1-13. Adequate voltage drive must be supplied from the audio source to obtain maximum output. Note that there is no transformer and few components are needed.

As in any push-pull Class-AB or Class-B configuration there is some distortion at the low bias crossover point where the conduction changeover is being made between one transistor and the other. This crossover distortion is quite evident if the output of the amplifier is displayed on an oscilloscope screen. Elaborate biasing networks and/or the use of an operational amplifier driver can help in the reduction of this type of distortion. In Class-AB and Class-B configurations the higher the signal level, the lower is the distortion because the absolute value of the distortion remains constant regardless of signal level.

The output transistors are TIP29 and TIP30 types. The Radio Shack 2018 and 2026 types are replacements for these. A high-capacitance output capacitor is needed so as not to attenuate low-frequency components. The driver is a 2N3904 npn transistor.

A basic quasi-complementary plan is shown in Fig. 1-14. In this arrangement the driving transistors operate as a complementary pair, while the output transistors are npn types. Phase inversion to the push-pull configuration is supplied by the complementary pair.

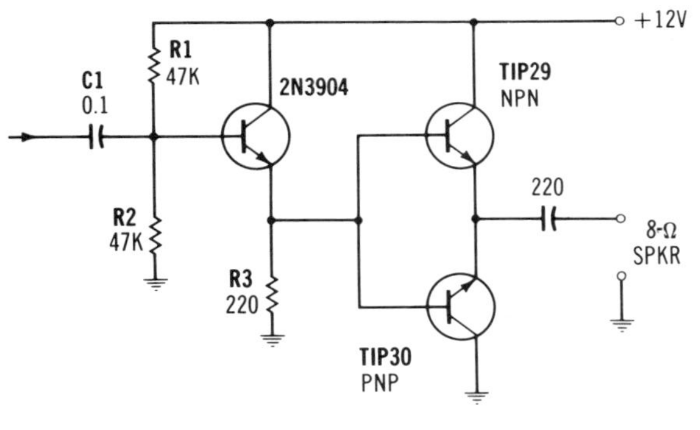

ALL RESISTORS ¼W

Fig. 1-13. Complementary-symmetry amplifier.

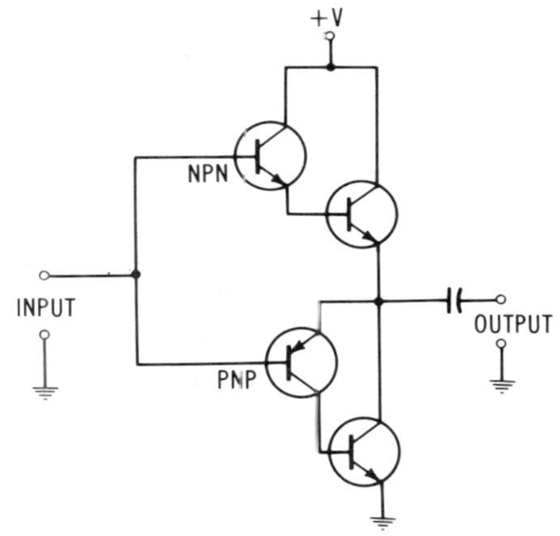

Fig. 1-14. Simplified schematic of quasi-complementary circuit.

FILE 1-9

PIERCE BIPOLAR CRYSTAL OSCILLATOR

The Pierce crystal oscillator has an untuned output, and the crystal is connected directly in the path between output and input. (See Fig. 1-15.) Crystal capacitance along with the interelement capacitances of the transistor and external capacitances establishes the Colpitts-like feedback arrangement that sustains oscillation. Capacitance *C1* in Fig. 1-15 is an aid in starting and maintaining oscillations at low frequency. The circuit will function well with fundamental-mode crystals extending over a range starting at 1.5 MHz and continuing above 15 MHz.

To start oscillations a bipolar oscillator circuit must be forward biased a small amount. This is the responsibility of resistors *R1* and *R2*. Resistor *R3* provides stabilization and a reasonably high base-to-emitter impedance, an aid to easy starting of the oscillator. A radio-frequency choke keeps the collector at high impedance but a low resistance for connecting the dc supply voltage to the collector. The Pierce crystal oscillator has a relatively high output impedance and will load circuits having impedance values beginning at 5000 ohms and extending upward. A lower-resistance load will reduce the output and, if too low, will stop the circuit from oscillating.

Oscillations can be checked by listening in on a receiver tuned to the crystal frequency or by displaying the output on the screen of a wideband oscilloscope. An alternative output indicator is the simple radio-frequency detector circuit of Fig. 1-16. This indicator can be built up and used as an indicator of rf output over a rather wide range of levels. It consists of a diode rectifier, X1, and appropriate resistor-capacitor filter. A dc meter is connected across the output. This can be a 0- to 50-microampere meter for very low level indications, or a 0- to 1-mA milliammeter for measuring higher levels of rf voltage. The diode can be a 1N34A, 1N914, or other general-purpose, low-level, silicon diode.

Sensitivity can be controlled with potentiometer *R3*. Circuit loading and additional attenuation of high-level signals can be accomplished with resistor *R1*. Although a value of 1 kilohm is indicated for *R1* it can be made as high in value as is needed.

In this circuit the signal is applied between input terminals 1 and 3 or between terminals 2 and 3. Also note optional resistor *R4*, which can be inserted when an appropriate load is to be placed on the output of the circuit under test. For example, this resistor might have an ohmic value of 50 ohms when the source of rf signal is to be matched to a 50-ohm load.

The output of the Pierce crystal oscillator is low level. Consequently, the output of the oscillator is applied between terminals 2 and 3 of the indicator. No resistor is used in the position indicated by *R4* because the Pierce circuit must be fed into a high-impedance load.

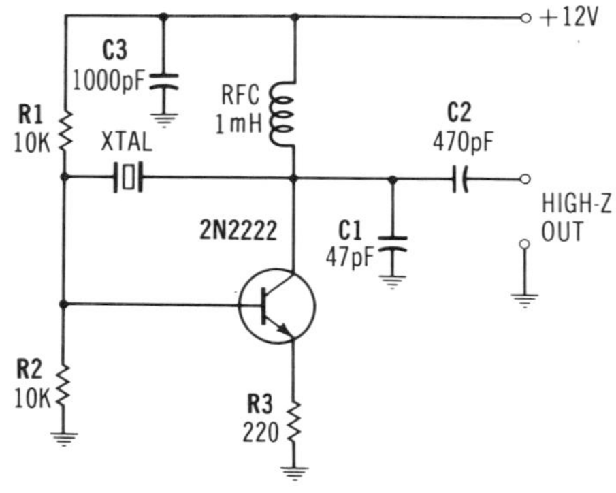

Fig. 1-15. Pierce bipolar crystal oscillator.

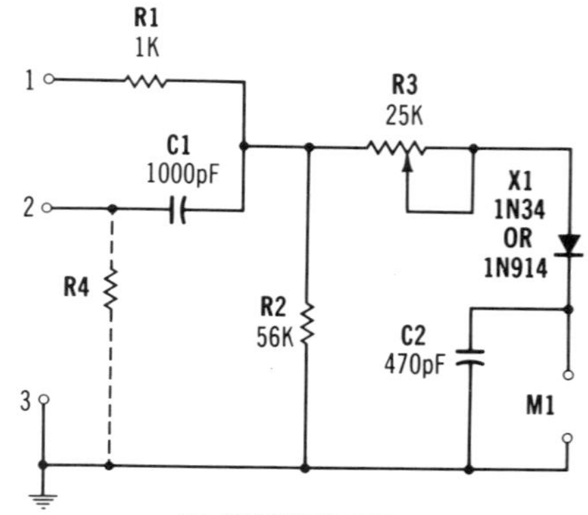

Fig. 1-16. Output indicator.

CRYSTAL OSCILLATOR WITH TUNED OUTPUT

An efficient crystal oscillator with a tuned output is shown in Fig. 1-17. The advantages of a tuned-output arrangement are a greater output and a capability for matching into a low-impedance load. The circuit is basically a Miller configuration with the interelement capacitances of the transistor serving as the feedback path. As in the case of the Pierce crystal oscillator or any bipolar crystal oscillator, some forward bias is required. This bias current is handled by resistors *R1* and *R2*. Resistor *R3* provides stability and limits the collector current to a safe value. Some additional output can be obtained by shunting an appropriate filter capacitor across resistor *R3*.

The circuit functions well with fundamental-mode crystals as well as third-harmonic types. In fact, even a fundamental-mode crystal can be operated in harmonic fashion by tuning the output resonant circuit to the desired harmonic frequency. For example, this adjustment would permit a 7-MHz crystal to develop a 21-MHz output if the resonant collector circuit is tuned to 21 MHz.

The preferred type of resonant circuit is a toroid core and associated variable capacitor. The author's arrangement was designed to operate on the 40-meter amateur band and used an Amidon T-50-2 toroid (red). It was made up of 35 turns of enameled No. 26 wire. The secondary winding consisted of 4 turns of No. 26 wire mounted over the center of the primary. This provided a suitable match into a 50-ohm load. The variable capacitor was an Arco 463 or J. W. Miller 160-D 180-pF compression trimmer capacitor.

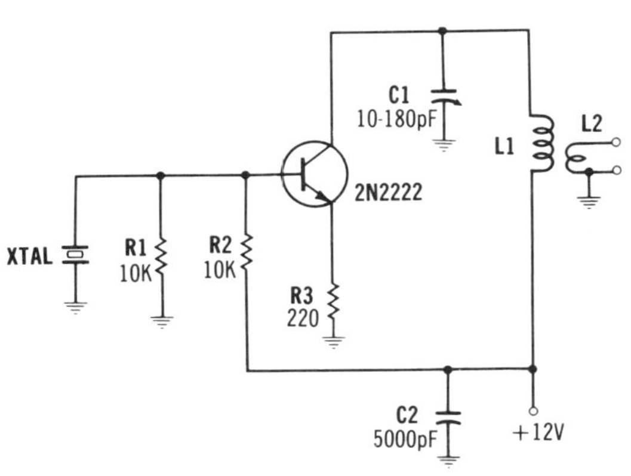

Fig. 1-17. Crystal oscillator with tuned output.

The output can be observed on a suitable wideband oscilloscope or picked up on a receiver tuned to the crystal frequency. The indicator of Fig. 1-16 can also be used to evaluate the output. In this example a 50-ohm resistor is connected between input terminals 2 and 3 with the secondary winding of the toroid transformer connected across the same two terminals.

Toroid coils for other bands and frequencies can be wound if you know the desired inductance and the proportionality constant of the core material. This constant (K) for the Amidon T50-2 (0.5 to 30 MHz) is 5.0, and for the T50-6 (3 to 250 MHz) is 4.0 The equation is:

$$L = KN^2$$

or

$$N = \sqrt{L/K}$$

where
L is in nanohenrys,
N is number of turns.

Coil data made available by Amidon Associates (12033 Otsego Street, North Hollywood, CA 91607) are of help if you wish to do some coil winding for the various radio amateur bands. The basic resonant frequency equation is

$$f = \frac{1}{2\pi\sqrt{LC}}$$

where
f is the frequency in hertz,
L is the inductance in henrys,
C is the capacitance in farads.

This equation can be simplified and solved for inductance values that are practical for the range of hf and vhf radio amateur bands:

$$L = \frac{25{,}330}{f^2C}$$

where
L is the inductance in microhenrys,
f is the frequency in megahertz,
C is the capacitance in picofarads.

A nomograph for simple parallel and series resonant circuits is given in Fig. 1-18. For ex-

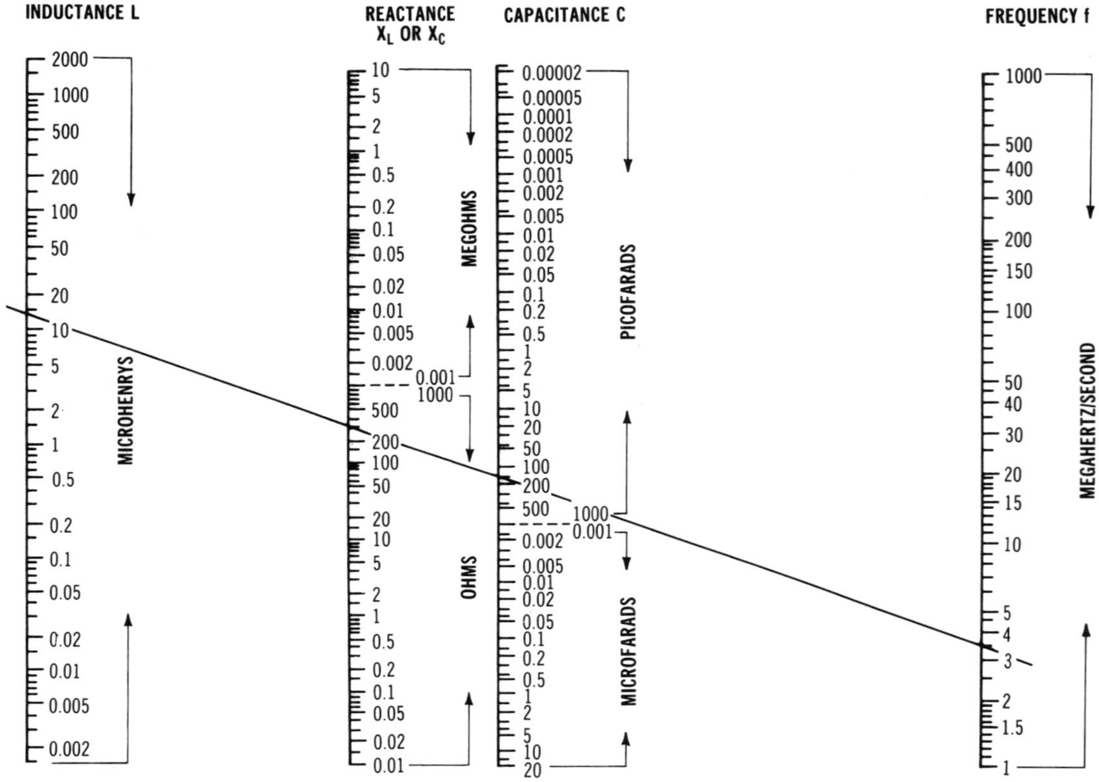

Fig. 1-18. ***LCR*** **nomograph.**

Table 1-1. Inductance of Toroid Coils as Related to Number of Turns of Closewound No. 24 Enameled Wire

E Type (Red) Toroid (Microhenrys)	Turns No. 24	SF Type (Yellow) Toroid (Microhenrys)
15.3	44	
14.1	42	
12.5	40	
11.5	38	
10.3	36	5.6
9.3	34	5.1
8.4	32	4.5
7.5	30	4.0
6.8	28	3.7
6.1	26	3.3
5.5	24	3.0
4.8	22	2.6
4.1	20	2.3
3.6	18	2.1
3.1	16	1.7
2.6	14	1.4
2.1	12	1.2
1.5	10	1.0

ample, if frequency and capacitance are known, a straightedge can be used to determine the inductance required to obtain resonance at the specific frequency. Table 1-1 presents the number of turns of No. 24 enameled wire required to obtain a specific inductance for two Amidon cores used widely for the high-frequency spectrum. The left column gives inductance values for the type E toroid, which is widely used for frequencies below 30 MHz. The third column gives similar data for the yellow SF core, which is recommended for frequencies above 10 MHz.

If a low-impedance output is to be obtained, an additional few-turns secondary must be wound on top of the primary. Remember that the transformer impedance ratio is equal to the square of the turns ratio, or $Z_P/Z_S = (N_P/N_S)^2$.

An example of use is given in Fig. 1-18. Assume a resonant frequency of 3.5 MHz and a capacitance of 150 pF. The required inductance is obtained by drawing a line through 3.5 MHz and

150 pF and continuing it to where it crosses the inductance column. In Fig. 1-18 this is at approximately 14 microhenrys. The next step is to determine the number of turns of No. 24 enameled wire, which, when closewound on the toroid, will produce the required inductance. Note in Table 1-1 that for an inductance of 14.1 microhenrys a total of 42 turns is required. This set of constants would obtain resonance at the low-frequency end of the 80-meter band. By varying the capacitance from high to low, the same combination would tune to the higher frequencies of the 80-meter band.

If you prefer to use a standard single-layer closewound coil on a 1.25-inch-diameter coil form, you will find the required information in Table 1-2. Data are given for the six high-frequency radio-amateur bands. Also, the recommended number of turns for a secondary winding mounted immediately under the primary winding is given. The actual number of secondary turns, of course, depends on the impedance ratio you wish to establish. The primary winding turns are based on using a 100-pF variable capacitor. Some adjustment of the number of primary winding turns may be necessary on the 160-meter band, depending on which end of the band is to be covered.

Table 1-2. Coil-Winding Data*

Meter Band	Primary Winding	Secondary Winding
160	60 turns, No. 26	8 turns, No. 20
80	45 turns, No. 22	6 turns, No. 20
40	21 turns, No. 22	4 turns, No. 20
20	11 turns, No. 22	3 turns, No. 20
15	8 turns, No. 20	2 turns, No. 20
10	5½ turns, No. 20	1½ turns, No. 20

*All turns closewound on 1¼-inch-diameter form using enameled wire.

FILE 1-11

BIPOLAR CLASS-C AMPLIFIER

Bipolar transistors can be operated in Class A, Class B or Class C. When a bipolar transistor is forward-biased Class A, the bias point is usually centered on the linear portion of the transfer characteristic (Fig. 1-19). In Class-B biasing the bias point is at collector current cutoff. The base-emitter junction is not biased or is reverse biased for Class-C operation. In practice a voltage of at least 0.7 volt between base and emitter is needed to obtain forward biasing. If no bias (zero bias) is applied, the bipolar transistor is automatically biased Class C. Consequently, the very simple circuit of Fig. 1-20 shows a practical and common method of biasing a bipolar Class-C rf power amplifier.

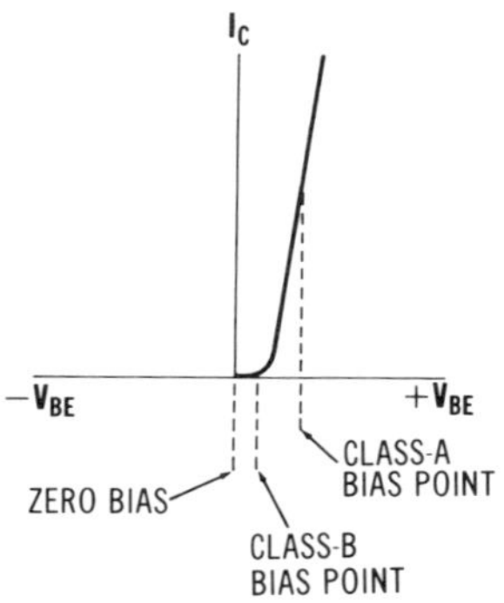

Fig. 1-19. Biasing of rf power amplifier.

The Class-C stage conducts collector current only during a portion of the positive alternation of the input signal. Usually the input signal is applied by way of a step-down transformer because of the very low input impedance of the bipolar device. The bursts of collector current excite the output resonant circuit and a sine wave of rf voltage develops across the output. A transformer arrangement is used to match this signal to a low-impedance load when a low-power transistor with a relatively high output impedance is used. In practical Class-C rf power amplifiers a more elaborate network is needed for matching because of the relatively low output impedance of a power bipolar transistor.

The low output impedance of the typical bipolar power transistor means that the circuit functions well as a wideband amplifier. A typical rf power amplifier for wideband application is shown in Fig. 1-21. The π network provides a match between a 50-ohm load and the low output impedance of the collector circuit. Such a network also attenuates harmonic components and higher-frequency spurious signal variations. In a practical stage the network Q can be chosen to provide a bandwidth of 0.5 to several megahertz, depending on the center frequency.

The source of signal must come from a low-impedance load. In addition a low-value base resistor is advisable. The input circuit should not be overdriven. These precautions prevent the reverse breakdown of the base-emitter junction, which can result in transistor damage.

In the circuit built by the author for 40-meter operation capacitors *C1* and *C2* had values of 470 pF. Inductor *L1* was a T-50-2 toroid of 14 turns of No. 22 enameled copper wire.

In a test procedure the low-impedance output of the crystal oscillator shown in Fig. 1-17 was applied to the low-impedance input of the amplifier. The low-impedance output of the amplifier was connected to the output indicator of Fig. 1-16 with the 50-ohm load resistor *R4* in place. A substantial amplification of the crystal oscillator signal was noted. The combination operates as a small QRP transmitter for 40 meters. One need only key the ground side of the oscillator. The amplifier can be made to operate on various amateur bands and other frequencies, too, by proper selection of π-network components *C1*, *C2*, and *L1*.

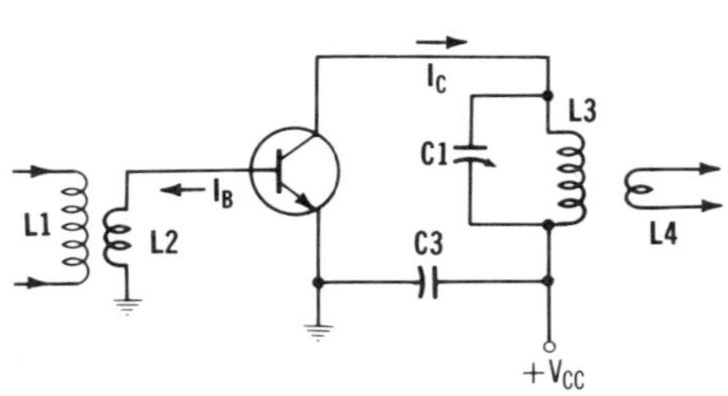

Fig. 1-20. Basic rf power amplifier.

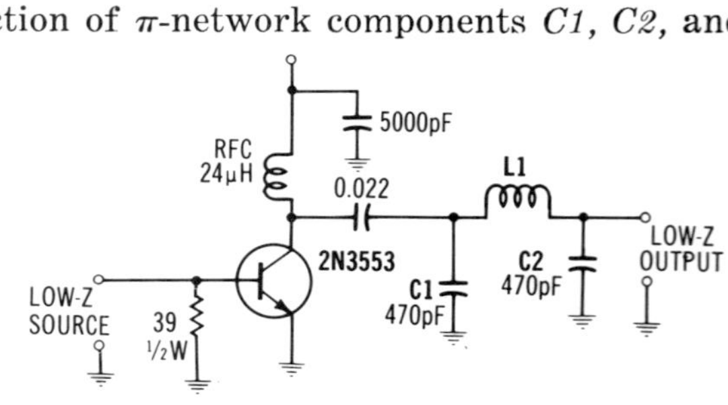

Fig. 1-21. Class-BC amplifier.

VARIABLE-FREQUENCY BIPOLAR OSCILLATOR

A variable-frequency oscillator (vfo), if suitably calibrated and stable, can be used in place of a crystal-controlled oscillator. Oscillators should not drift or jump in frequency. Their electrical design and mechanical rigidity are factors that influence stability. A well-regulated power supply and a high-stability oscillator circuit are essential. Select the parts well and employ a firm and rigid construction. The variable capacitor that tunes the oscillator should be of good quality and have an isolated tuning shaft. Proper shielding is essential in reducing stray capacitance effects.

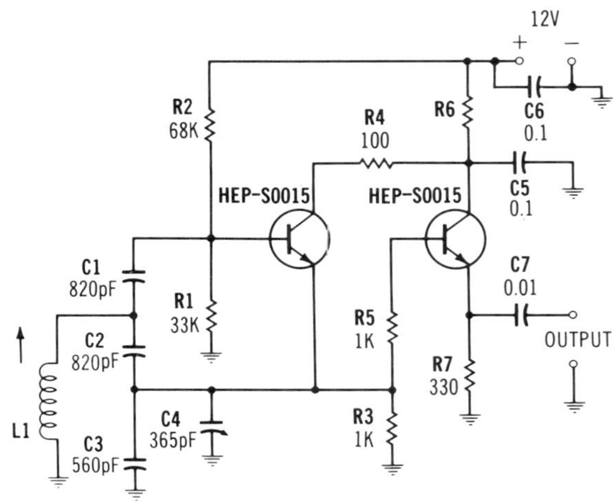

Fig. 1-22. Variable-frequency oscillator.

The circuit shown in Fig. 1-12 is a Seiler-type oscillator using an inductor, *L1*, that permits operation on the 40-meter band. The inductor is adjustable between 0.735 and 0.984 μH. It is a J. W. Miller 20A827RB1. Other similar coils can be used, or you can wind a toroid and adjust the number of its turns to obtain correct tuning over the 40-meter band. In adjusting the oscillator the inductor, *L1*, is preset in such a manner that variable capacitor *C4* is sure to accommodate the required tuning range of 7.0 to 7.3 MHz. For best stability, capacitors *C1*, *C2*, and *C3* should be polystyrene types.

The oscillator proper is followed by an emitter-follower stage that provides isolation and light loading of the oscillator as well as a low-impedance output. The transistor used in the author's version was a Motorola HEP-S0015. Other small-signal, general-purpose types will function in the circuit. The indicator of Fig. 1-16 will respond to the output level of the vfo.

If you wish to tune finely over a portion of the band, an additional 15-pF variable capacitor can be connected in parallel across the main-tuning 365-pF variable capacitor.

FILE 1-13

ASTABLE MULTIVIBRATOR

A multivibrator oscillator is a two-transistor configuration that generates a square-wave or a pulse output. It is in effect a two-stage amplifier fed back upon itself, with the collector of each stage linked back to the base of the other stage (Fig. 1-23). The feedback is strong, however, and the two transistors are alternately driven beyond cutoff. During one alternation the charge built up on capacitor *C1* holds the first transistor at cutoff. The cutoff interval is determined by the time constant of resistor *R1* and capacitor *C1*. When the first transistor returns to conduction the second one is driven to the cutoff region. Now it is held cut off by the time constant contributed by resistor *R2* and capacitor *C2*. If the two time constants are equal and if the values of resistors *R3* and *R4* are low enough, a good square wave is developed at the output. This square wave can be obtained at either collector of the multivibrator.

If an oscilloscope is available, the square wave can be observed. For the constants shown in Fig. 1-23 the operating frequency falls between 800 and 1000 hertz. If the square-wave frequency is to be increased, it is only necessary to lower the two time constants, *R1C1* and *R2C2*, a like amount. A lower frequency can be obtained by increasing these time constants.

An asymmetrical or pulselike wave can be obtained by using unequal time constants. For example, a positive pulse can be obtained by decreasing the *R2C2* time constant relative to the *R1C1* value. This will ensure that the second transistor is cut off for a shorter period than the first transistor. It is during this interval that the voltage is positive on the collector of the second transistor and, therefore, a positive pulse appears at the output. If desired, a negative pulse can be obtained at the same time from the collector circuit of the first transistor.

If the frequency is to remain the same but a pulse output is to be obtained, it is necessary to increase time constant *R1C1* by the same amount that the time constant of *R2C2* is decreased. The duration of the positive pulse in the output depends on how much shorter time constant *R2C2* is, compared with time constant *R1C1*.

A multivibrator of this type is a simple circuit for obtaining a square-wave or pulse output. Furthermore, the frequency or duration of the square wave or pulse can be changed conveniently. Any general-purpose bipolar transistor will function well in the circuit.

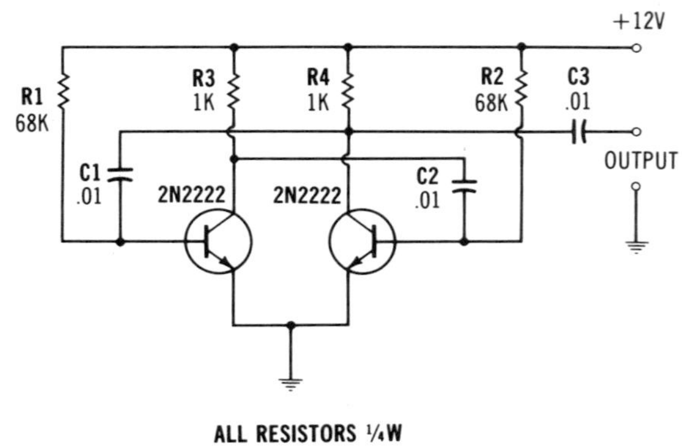

Fig. 1-23. Astable multivibrator.

UNIJUNCTION TRANSISTOR OSCILLATOR

A unijunction transistor functions well as an oscillator. It has the ability to generate very low frequency pulses of very short duration. Of course, it can also be used to provide higher-frequency outputs. It is noted, however, for generating very short pulses that are used in triggering applications. Such a unijunction oscillator (Radio Shack 276-2029 or similar type) is shown in Fig. 1-24.

When the circuit is turned on, capacitor *C1* charges toward a positive value through resistor *R1*. This charging action finally supplies a potential to the emitter above the trigger value. Consequently, the unijunction goes into a high-current state. The impedance, of course, is now low between emitter and base 1. Thus the capacitor is discharged quickly, developing a sharp positive pulse across resistor *R3*. This is the positive pulse output of the oscillator. The emitter voltage now reverse biases the transistor and the output ceases. Then capacitor *C1* again begins to charge, eventually reachinig the conduction point again to initiate a second output pulse.

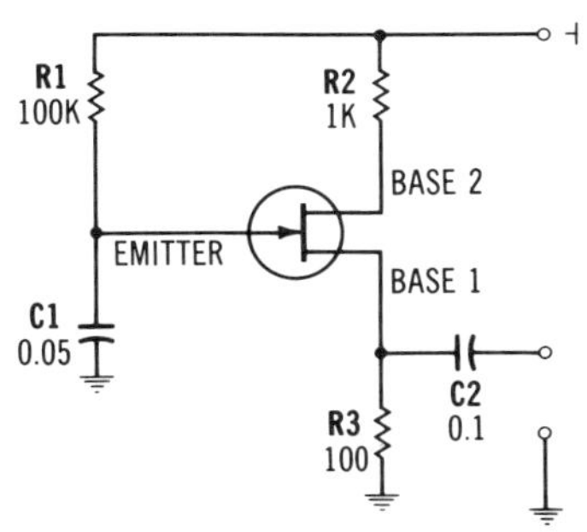

Fig. 1-24. Unijunction oscillator.

The frequency of the unijunction oscillator is determined largely by the time constant of resistor *R1* and capacitor *C1*. This time constant is made very long if a low-frequency pulse train is to be obtained. Lowering the time constant will increase the output frequency. If an adjustable frequency output is desired, resistor *R1* can be a fixed resistor and a potentiometer. A practical combination for replacement of resistor *R1* would be a 10-kilohm fixed resistor in series with a 1-megohm potentiometer.

FILE 1-15

LED DRIVER

A bipolar transistor, such as the general-purpose 2N2222, can be used for driving a light-emitting diode (LED). In this application the circuit serves as a high-impedance voltage indicator. In many applications where an LED indication would be desirable the low impedance of the LED loads down the circuit point under test. An intermediate transistor can be used to place a substantially higher impedance load on the circuit that is to illuminate the LED when a specific voltage is present. At the same time, the current amplification of the intermediate transistor draws a significant current through the LED to cause it to illuminate.

A typical circuit is given in Fig. 1-25. Note that a high-value resistor is inserted between the point of excitation and the base of the transistor. Therefore there is a very low base current and an insignificant load is place on the voltage source. However, the resultant base current when a positive voltage is applied to the left side of resistor *R1* results in a significant collector current. This current is limited by resistor *R2* but is sufficient to cause LED illumination.

The circuit of Fig. 1-25 is a very basic logic indicator. When logic 1 (+5 volts) is applied to the input, the LED glows. If a logic 0 (ground) is applied to the input, the LED will be dark.

Check out the circuit of Fig. 1-25 using a 6-volt battery. Note that for logic 1 the LED is lit; for logic 0 it is not lit. The values of resistors *R1* and *R2* also control the illumination. If you shunt a 180-ohm resistor across resistor *R2*, there will be some increase in brightness. However, adequate illumination is obtained with the 270-ohm value selected and a lesser load is placed on the power supply. Some slight increase in brightness can also be obtained by decreasing the value of resistor *R1*. A lower value of resistor *R1*, however, can place more load on the circuit that is to cause LED illumination.

The average LED will produce excellent illumination when current is 15 to 20 milliamperes although satisfactory illumination can be obtained with a current as low as 5 milliamperes. For most types the absolute maximum current should not exceed 50 milliamperes.

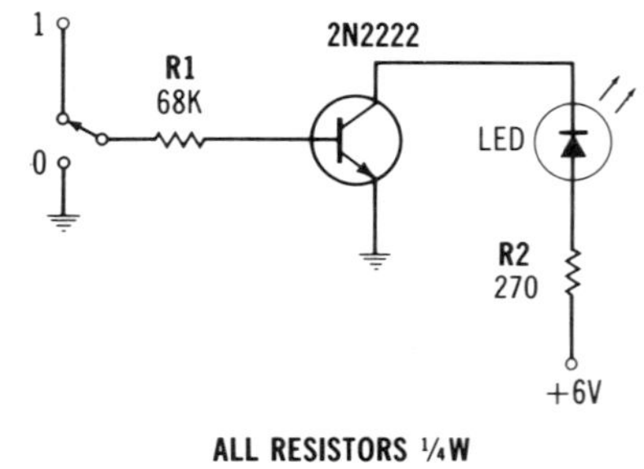

Fig. 1-25. LED driver.

FILE 2

FET CIRCUITS

FET COMMON-SOURCE AMPLIFIER

A field-effect transistor is a three-element device consisting of source, gate and drain (Fig. 2-1). It is basically a single-junction (unipolar) device made up of a semiconductor bar (n-type material in the example) into which two facing strips of semiconductor material of the opposite type are diffused.

In FET terminology the bar is called a *channel* while the element that controls the motion of charges along the channel is known as the *gate*. The common end of the channel is known as the *source* while the opposite end is called the *drain*. When the semiconductor bar is biased as in Fig. 2-1, there is a motion of electrons from source to drain through the bar or channel. In the fundamental operation of the FET the gate is able to control this motion of charges along the channel. Let is consider how this is accomplished.

In a practical circuit the gate-to-channel junction is reverse biased, with the source being used as the common connection (Fig. 2-2). This reverse biasing of the junction causes a depletion region to extend into the channel. The presence of the depletion region in the channel causes an increase in the resistivity of the bar (decrease in conductivity). When the gate bias is increased there is a further rise in the channel resistance and the motion of charges (drain current) is reduced. If the gate bias voltage is made high enough, the channel or drain current can be reduced to zero, producing cutoff.

The basic operation of a common-source FET amplifier along with a typical circuit and the transfer characteristic is given in Fig. 2-3. An applied audio or other signal causes the gate voltage to vary about the gate bias set by the gate bias voltage. As the gate voltage varies so does the depletion region vary back and forth in the channel, causing a similar change in the channel resistance. Consequently the gate voltage variation causes a similar variation in the channel or drain current. In fact, as the gate voltage is made to vary with signal there results a substantial change in the drain current. The actual drain output voltage results from the variation of the drain current in the drain load resistor. The actual voltage gain of the stage is, of course, the ratio V_o/V_{in}.

A positive swing of the gate voltage produces an increase in the drain current. In turn, there is a drop in the drain voltage (the output voltage swings negative). The negative alternation of the input voltage decreases the drain current and there is a positive swing of the drain and output voltage. The input and output voltages are said to be of opposite polarity, or 180° out of phase.

In the example the channel was composed of n-type material and therefore the charge carriers (current) were electrons. The gate has many holes, or positive charge carriers, which neutralize some of the electron carriers to produce the depletion region that controls the resistance of the channel. This type of FET is known as an n-type junction field-effect transistor.

There are also p-type junction FETs. In this type the current in the channel is the result of the motion of hole charges. The gate has electron carriers and they in turn neutralize the hole carriers to form the depletion region that moves in and out of the channel in accordance with gate voltage change. The operation is still the same, with the

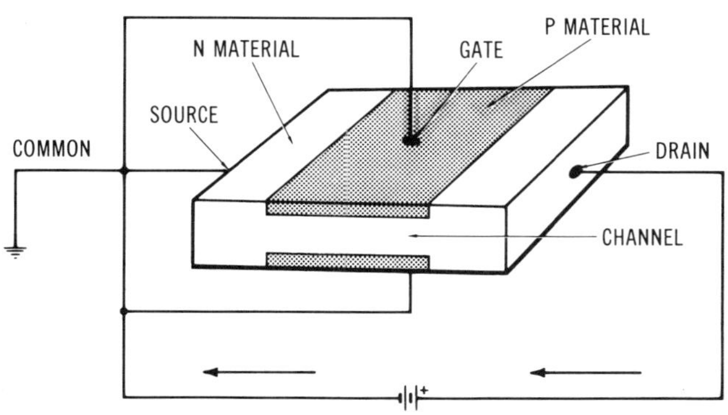

Fig. 2-1. Basic FET construction.

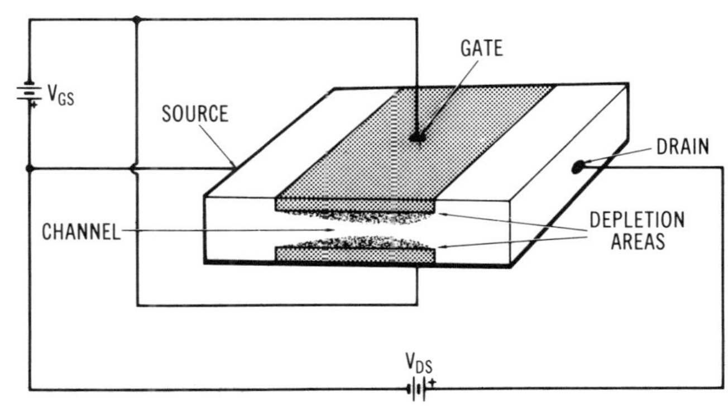

Fig. 2-2. Gate biasing forms depletion areas.

channel current being a function of the gate voltage. The only difference is that the drain voltage is made negative instead of positive as in the case of the n-type junction FET.

The field-effect transistor is basically a voltage amplifier rather than a current amplifier, which the bipolar transistor is. There is no significant gate current, and therefore the input displays a high impedance to any signal source. The gate circuit responds to an input voltage change rather than an input current change. The output impedance, too, is moderately high and is higher than that displayed by the usual bipolar transistor.

The practical common-source FET amplifier is given in Fig. 2-4. No external gate bias is necessary. Rather the presence of drain current in source resistor *R2* develops the negative gate-to-source dc bias. Resistor *R1* can have a high ohmic value because there is no significant gate current; this, too, aids in keeping the input impedance high. Resistor *R2* is kept low in value so as to develop the voltage that will bias the transistor stage on the linear portion of its transfer characteristic. Typical values are given for a Radio Shack 276-2035 general-purpose, small-signal FET. FETs with similar characteristics will operate well in the circuit. It may be necessary to make an adjustment in the ohmic value of resistor *R2* to establish linear biasing for other types.

Resistor *R3* is the drain load resistor across which the output signal is developed. Its value is selected to maintain the proper dc collector voltage and collector current.

Capacitor *C2* prevents an ac voltage from developing across the source resistor *R2*. Such an ac voltage would be degenerative and reduce the amplifier gain.

An oscilloscope is an aid in finding precise resistor values for the amplifier if you wish to obtain the greatest output voltage for a given input signal voltage. An oscilloscope connected across the output will show if the input signal of a given value produces a symmetrical, nondistorted output. Adjustments can then be made in the ohmic value of resistor *R2* to obtain correct biasing for desired operating conditions. Resistor *R3* can also be adjusted in value to obtain maximum output. Usually it can be of higher value for a very weak input signal which is to be amplified a maximum amount. The values given in Fig. 2-4 are for good general-purpose operation of the stage. These values produce an operating dc drain voltage of 6 volts.

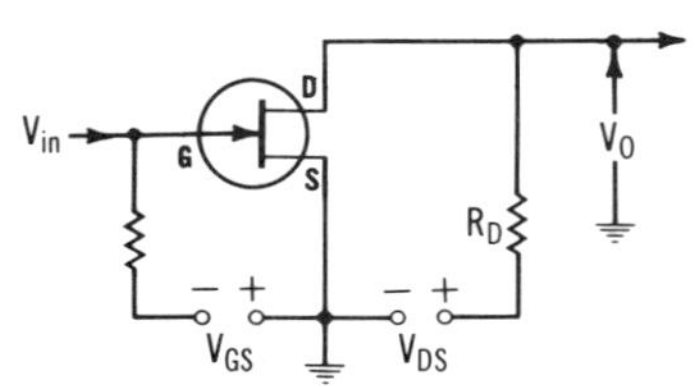

(A) Common-source circuit.

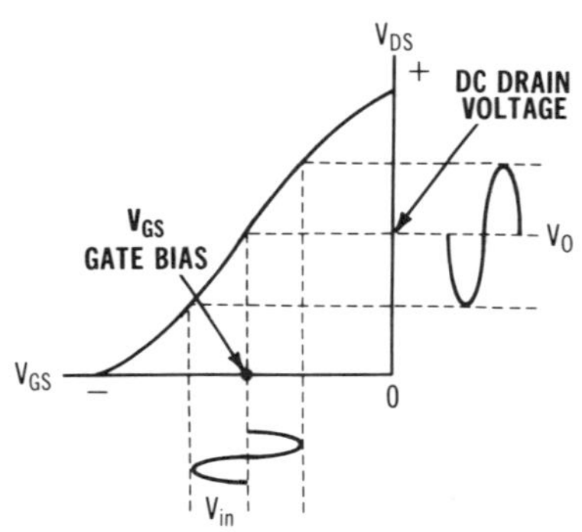

(B) Transfer characteristic.

Fig. 2-3. Basic FET operation.

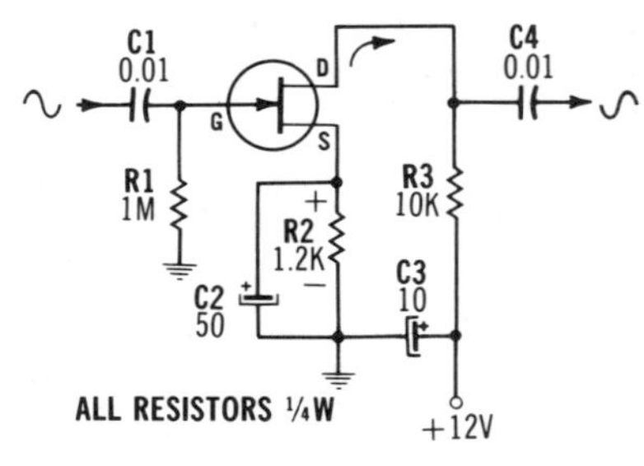

Fig. 2-4. Basic common-source amplifier.

COMMON-GATE AMPLIFIER

In the common-gate configuration (Fig. 2-5) a signal is applied between the source and common and removed between the drain and common. The File 2-1 circuit was that of a common-source stage with the signal being applied to the gate and removed at the drain. The common-gate amplifier is capable of voltage gain but has a current gain that is less than unity. The common-source stage is capable of both voltage gain and current gain.

The common-gate stage has a very low input impedance and a very much higher output impedance. Therefore it can be used in any circuit where there must be a transformation between low input impedance and high output impedance at the same time there is a possible voltage gain. It is not a particularly common FET circuit although it does have some application in radio-frequency amplifiers.

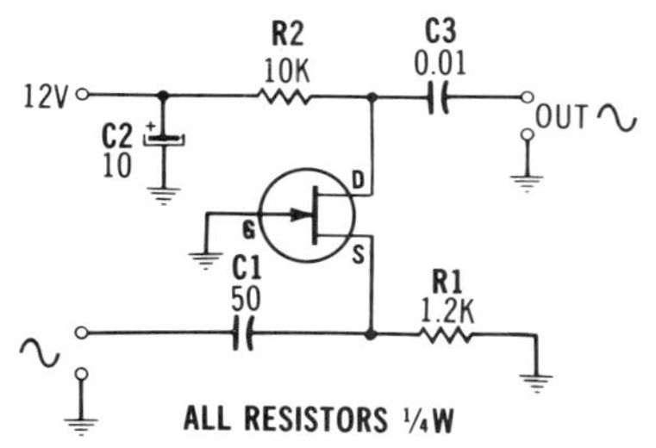

Fig. 2-5. Common-gate amplifier.

The Radio Shack 276-2035 field-effect transistor is used in the practical circuit of Fig. 2-5. Other general-purpose n-channel FETs can be used in the circuit. The component values are approximately the same as those used in the common-source stage of Fig. 2-4. No gate resistor is used. Instead, the gate is returned to the common. Signal is applied across the source resistor, and this resistance, along with the input resistance of the source, circuit presents a low impedance to the source of signal. It is necessary to increase the value of the input coupling capacitor because of the much lower input impedance of the common-gate stage. This increase maintains the frequency response to a low limit of at least 100 hertz. The high-frequency response is well over 20,000 hertz.

In operation the common-gate configuration input and output voltages are in phase. In the previous common-source circuit input and output voltages were 180° related.

In testing the circuit it is usually necessary to apply a higher input voltage than in the common-source circuit because of the loading effect of the low-impedance source-input circuit. The voltage gain is approximately the same as that obtained with the common-source stage. Maximum undistorted output can be obtained by proper regulation of the *R1* and *R2* resistor values.

FILE 2-3

COMMON-DRAIN (SOURCE-FOLLOWER) STAGE

In the common-drain configuration the input signal is applied to the base, and the output is removed from the emitter (Fig. 2-6). The common-drain, or source-follower, stage is an effective impedance transformer. The input impedance is very high, the output impedance quite low. Thus the stage is capable of current gain but not voltage gain. In a properly designed stage the output voltage is just slightly less than the input voltage. Such a stage is used mainly in matching a high-impedance source of signal to a low-impedance load. It does so with a minimum loading effect on the source of signal.

The output voltage is in phase with the input voltage. Another advantage of the common-drain stage is its capability of handling a higher input signal level without distortion while, at the same time, it places a very minimum load on the source of the signal.

The practical source-follower stage of Fig. 2-6 uses the Radio Shack 276-2035 n-channel field-effect transistor. Similar general-purpose FETs can be used in the circuit. The component values are approximately the same as used in the File 2-1 and File 2-2 schematics except that they are rearranged. Resistor *R1* in the gate circuit is returned to the junction of resistors *R2* and *R3*. Therefore proper gate bias is applied to the stage and is that voltage developed across resistor *R2*. Resistors *R2* and *R3* in combination establish the proper drain voltage and serve as the output load resistors of the circuit.

The stage will operate over a considerable range of audio input signal levels and has a low-frequency response that extends down to at least 100 hertz. The high-frequency response extends well above 20,000 hertz.

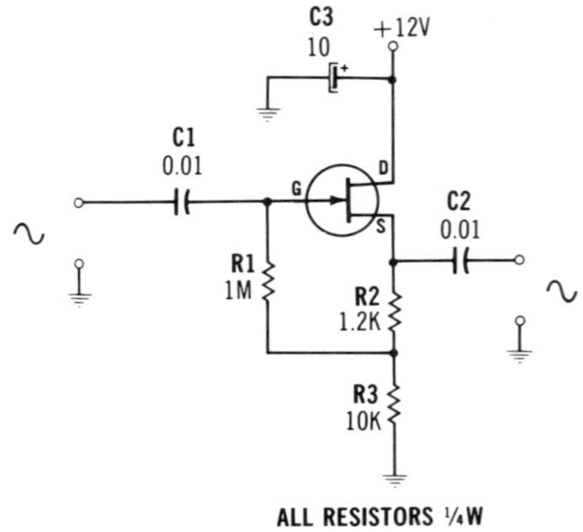

Fig. 2-6. Common-drain stage.

STABILIZED COMMON-SOURCE AMPLIFIER

Two factors have a significant influence on FET bias requirements. One is the production spread, which may be such that the saturation current I_{DSS} may have a ratio of 3 to 1 or higher. Temperature drift is the second factor and involves both the warm-up operating-point drift and a shift as the result of change in ambient temperature.

The changes in operating conditions as the result of production factors and FET temperature are pronounced when only external gate bias is used. Self-biasing is much more self-adjusting because any shift in operating-point drain current is compensated for by the change in bias produced by the change of current in the source resistor. Additional operating-point stabilization can be obtained with a further increase in the ohmic value of the source resistor. Normally this would result in an improper bias, but a combination of gate-divider positive bias and a higher value of source bias can be used to set up the required operating point. Refer to Fig. 2-7.

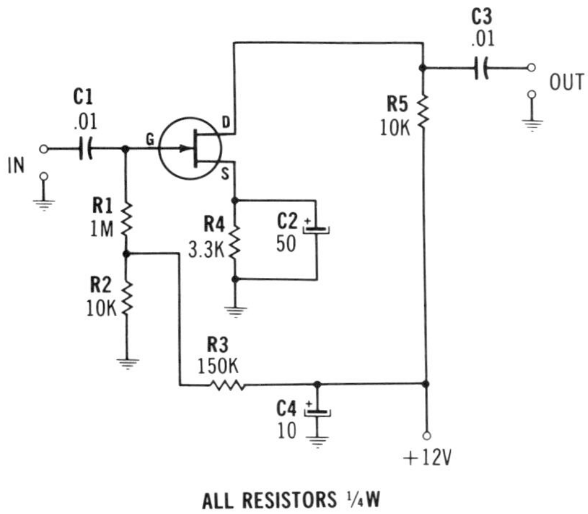

Fig. 2-7. Stabilized common-source amplifier.

Resistor *R4* is increased in value to improve stabilization. Its ohmic value, however, is too high for obtaining the correct operating bias. Resistors *R2* and *R3* are added to make appropriate compensation and the combination of the three resistors, *R2*, *R3*, and *R4* establishes the correct and stabilized operating bias for the stage. Good stabilization is obtained by increasing the ohmic value of the source resistance three to five times. The ratio of the two resistances *R2* and *R3* must be set to establish correct bias.

Resistors *R1* (gate resistor) and *R5* (drain resistor) are the same as those recommended for Fig. 2-4, when the Radio Shack 276-2035 field-effect transistor is used.

FILE 2-5

PHASE SPLITTER

The simple single-FET circuit of Fig. 2-8 provides a method of obtaining equal-amplitude but opposite-polarity output waves. Such a circuit can provide voltage drive to some form of balanced input, such as the input to a push-pull amplifier.

The output of the drain circuit is opposite in polarity to the input wave, while the source output circuit is in phase with the input. The amplitudes of both outputs are slightly less than the amplitude of the input wave. The Class-A linear bias is developed across source resistor *R2*. The utmost in output voltage level can be obtained by using the source filter capacitor, *C2*. In the interest of simplicity this capacitor can be eliminated if somewhat less output voltage is adequate.

Equal-amplitude but opposite-polarity outputs are developed across drain resistor *R3* and source resistor *R4*. The FET is the Radio Shack 276-2035. Other general-purpose transistors can be used, although it may be necessary to adjust the ohmic value of resistor *R2* to obtain Class-A linear biasing.

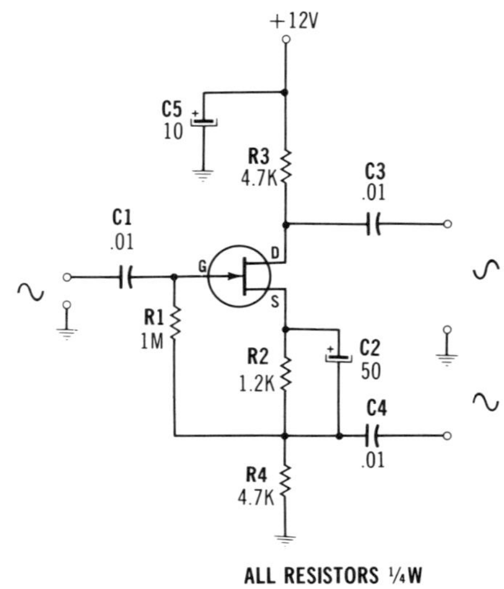

Fig. 2-8. Phase-splitter circuit.

PUSH-PULL AUDIO POWER AMPLIFIER

The push-pull connection provides a convenient way of increasing power output. The FET, because of its high input impedance, can be driven readily from a simple phase-splitter, such as that covered in File 2-5. Only a few additional components are needed to set up the push-pull power output arrangement of Fig. 2-9. The input stage is the phase-splitter circuit and develops the equal-amplitude and out-of-phase voltages that drive the opposite gates of the FET. In the circuit arrangement the phase splitter was the Radio Shack 276-2035, while the output FETs were the Siliconix 1.8-watt 2N4856s.

A miniature audio transformer matches the push-pull drain circuit of the output FET to an 8-ohm speaker. Use a transformer (Radio Shack 273-1380) with a primary winding impedance of 1000 ohms, center-tapped, and a secondary winding impedance of 8 ohms. Supply a 12-volt voltage connection to the center tap.

The values shown provide good low-frequency response down to several hundred hertz. Capacitor *C6* across the primary of the output transformer can be used to regulate the high-frequency response. The value shown gives good response in excess of 10,000 hertz. If less output is desired, capacitor *C2* in the phase-splitter circuit can be removed. A higher-power output can be obtained by increasing the supply voltage to 18 volts. Indeed, higher output can be obtained with even higher voltage. In such a case a larger output transformer and an increase in the wattage rating of resistor *R7* are recommended.

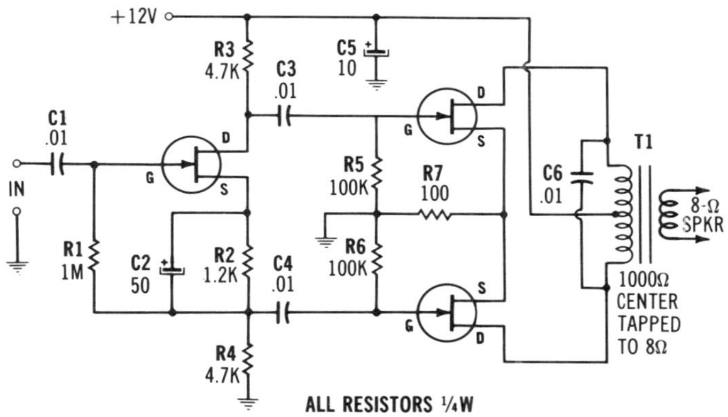

Fig. 2-9. Push-pull amplifier with phase-splitter circuit.

FET 100-kHz CRYSTAL OSCILLATOR

The 100-kHz oscillator is popular in calibrator circuits and as a source of signal for operating a chain of low-frequency dividers. The FET circuit of Fig. 2-10 performs well, and its waveform has a high harmonic content and provides useful calibration marks each 100 kHz up to 20–25 MHz and higher.

The fundamental frequency of operation is 100 kHz, but its harmonics can be tuned in on one of the WWV frequencies (2.5, 5, 10, 15, and 20 MHz). Trimmer capacitor *C1* can be adjusted to obtain an exact zero beat, setting the crystal oscillator precisely to a frequency of 100 kHz.

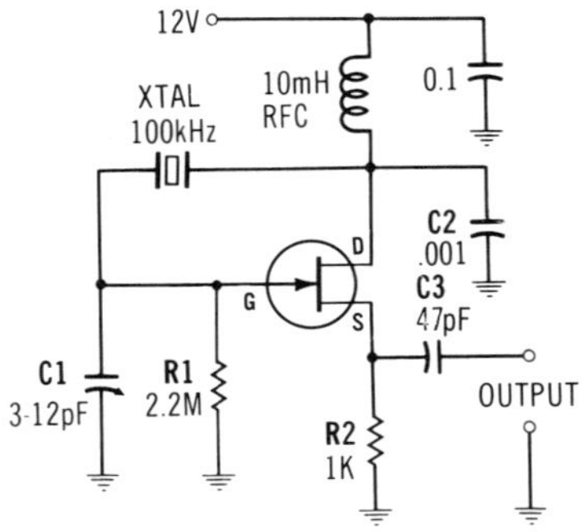

Fig. 2-10. FET 100-kHz crystal oscillator.

The output is removed from the source circuit at a relatively low impedance. The small value of output coupling capacitor *C3* provides greater attenuation at lower frequencies than at higher ones. This technique provides some emphasis of the higher-order harmonics without causing overload conditions by the stronger and much lower frequency harmonics and fundamental frequency.

A 1-MHz crystal also performs well in the circuit. Again, useful harmonics are available up to 20 MHz and higher. Calibration can again be made on one of the WWV frequencies of 5, 10, 15, or 20 MHz.

A Radio Shack 276-2035 FET was used in the sample circuit. Operation is possible with a 2N3819, MPF 102, or other small-signal, general-purpose, n-channel FET.

PIERCE FET CRYSTAL OSCILLATOR

The Pierce crystal oscillator has an untuned output, and the crystal is connected directly in the path between output and input circuits (Fig. 2-11). The crystal capacitance along with the interelement capacitances of the transistor establishes the Colpitts-like feedback arrangement that sustains oscillations. Very few components are needed in this simple oscillating circuit. The circuit functions well with fundamental-mode crystals, extending over a range starting at 1.5 MHz and continuing above 15 MHz.

Unlike the bipolar crystal oscillator, no forward biasing resistor combination is required. Resistor *R1* provides a gate-to-common current path, while resistor *R2* provides stabilization and limits the drain current to a safe value. A radio-frequency choke keeps the drain at high impedance at the same time the dc supply voltage is applied to the drain. The Pierce crystal oscillator has a relatively high output impedance and will load circuits having impedance values beginning at 5000 ohms and extending upward. A lower-resistance load will reduce the output and, if too low, it will stop the circuit from oscillating.

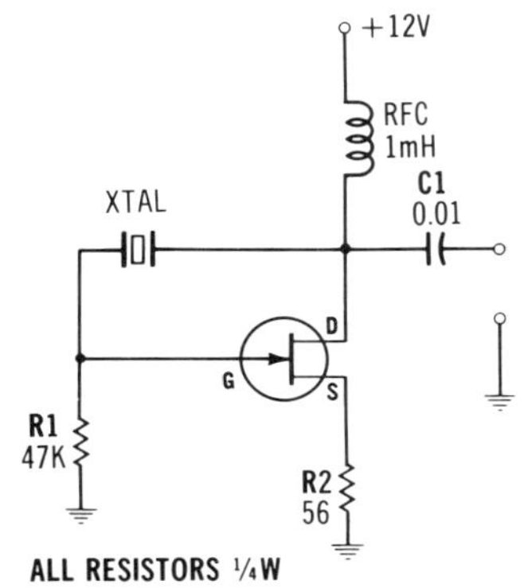

Fig. 2-11. FET Pierce crystal oscillator.

Oscillations can be checked by listening in on a receiver tuned to the crystal frequency or by displaying the output on the screen of a wideband oscilloscope. An alternative output indicator is the simple radio-frequency detector circuit of Fig. 1-16. This indicator can be built up and used as an indicator of rf output over a rather wide range of levels. Complete information on its operation is given under File 1-9. A variety of general-purpose FETs will operate in the crystal circuit, including the MPF 102 and the Radio Shack 276-2035. Two big advantages of the Pierce crystal circuit are that no output resonant circuit is required and a shift in frequency requires only that one change the crystal.

FILE 2-9

FET TUNED-OUTPUT CRYSTAL OSCILLATOR

An efficient crystal oscillator with a tuned output and only a few circuit components is shown in Fig. 2-12. The advantages of a tuned-output arrangement are greater output and the capability for matching into a low-impedance load. The circuit is basically a Miller configuration with the interelement capacitances of the FET serving as the feedback path. As in the case of the Pierce FET crystal oscillator, no forward biasing is required. Only a return-path resistor, *R1*, is required in the gate circuit. Resistor *R2* provides stability and limits the drain current to a safe value. Some slight additional output can be obtained by shunting an appropriate filter capacitor across resistor *R2*.

The circuit functions well with fundamental-mode crystals as well as third-harmonic types. In fact, even a fundamental-mode crystal can be operated in harmonic fashion by tuning the output resonant circuit to the desired harmonic frequency. For example, this arrangement would permit a 7-MHz crystal to develop a 21-MHz output if the resonant drain circuit is tuned to 21 MHz.

The preferred type of resonant circuit is a toroid-core transformer and associated variable capacitor. The author's sample was operated on the 40-meter amateur band and used an Amidon T-50-2 toroid (red). It was made up of 35 turns of No. 26 enameled wire. The secondary winding consisted of 4 turns of No. 26 mounted over the center of the primary. This provides a suitable match into a 50-ohm load. The variable capacitor was an Arco 463 180-pF compression trimmer.

The output can be observed on a suitable wideband oscilloscope or picked up on a receiver tuned to the crystal frequency. The indicator of Fig. 1-16 can also be used to evaluate the output. In this example a 50-ohm resistor should be connected between input terminals 2 and 3 of the indicator, with the secondary winding of the toroid transformer connected across the same two terminals.

Toroid coils for other bands and frequencies can be wound if you know the desired inductance and the core proportionality constant of the core material. This constant (K) for the Amidon T-50-2 (0.5 to 30 MHz) is 5.0 and the T-50-6 (3 to 250 MHz) is 4.0. The equation is:

$$L = KN^2$$

where
L is the inductance in nanohenrys,
N is the number of turns.

The equation for the number of turns is simply

$$N = \sqrt{L/K}$$

Small-signal general-purpose FETs, such as the MPF 102 and Radio Shack 276-2035, function well in the circuit. If additional power is required, use a Siliconix 2N4856; this FET will produce a greater power output. In fact, the power level is adequate for driving the bipolar Class-C amplifier covered in File 1-11.

Additional details on coil winding are given in File 1-10. Toroid and regular coil types are covered there.

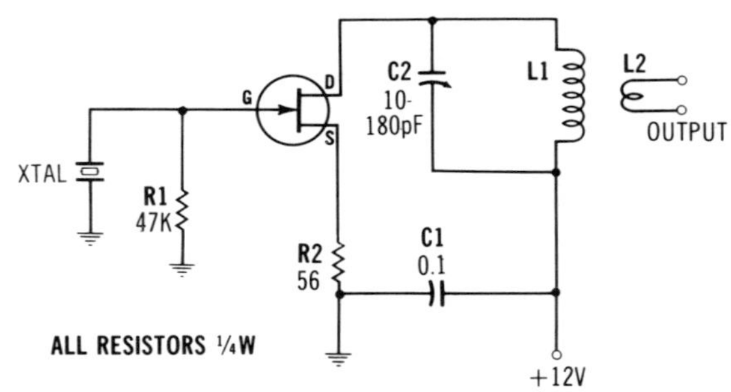

Fig. 2-12. Miller crystal oscillator.

FET RF OSCILLATOR-AMPLIFIER

A simple two-stage oscillator-amplifier that can be used as a miniwatt transmitter for the 40-meter band is given in Fig. 2-13. The first stage operates as a Pierce crystal oscillator and supplies drive signal to a simple FET Class-C amplifier. When driven by the output of the crystal oscillator, the small amount of gate current in resistor *R3* develops a bias on capacitor *C1*. The bursts of drain current excite the drain resonant circuit, developing an amplified rf voltage.

A radio-frequency output at low impedance is removed from across the several-turn output winding. A toroid coil and associated variable capacitor form the drain resonant circuit. Designed to operate on the 40-meter amateur band, an Amidon T-50-2 toroid was wound with 35 turns of enameled No. 26 wire. The secondary winding consisted of 4 turns of No. 26 mounted into a 50-ohm load. The variable capacitor was an Arco 463 180-pF compression trimmer. Data for regular and toroid coils for other bands and frequencies are given in File 1-10.

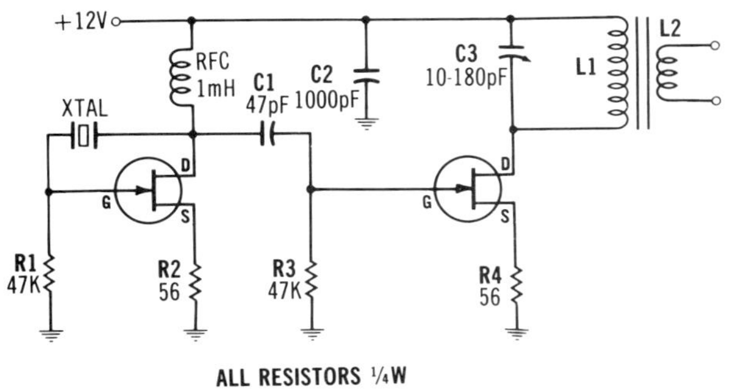

Fig. 2-13. FET oscillator-amplifier.

The output can be observed on a suitable wideband oscilloscope or picked up on a receiver tuned to the crystal frequency. The indicator of Fig. 1-16 can also be used to evaluate the output. In this case be sure to place an appropriate 50-ohm termination on the output.

The FETs used in the oscillator-amplifier can be MPF 102, 276-2035, or other low-signal, general-purpose types. In the author's example a higher-powered output FET, a Siliconix 2N4856, for the amplifier, provided a stronger output and received adequate drive from a Radio Shack 276-2035 used in the Pierce circuit.

FILE 2-11

PUSH-PULL CRYSTAL OSCILLATOR

A boost in crystal oscillator output can be obtained by connecting two field-effect transistors in push-pull (Fig. 2-14). The MPF 102, Radio Shack 276-2035, and other general-purpose types will oscillate strongly in this circuit.

Fundamental-mode and harmonic-frequency crystals will operate over a wide frequency range when used with an appropriate tuned-output resonant circuit. Coil data for standard or toroid forms can be obtained by referring to File 1-10. The only difference in the coil construction is the need for a center tap on the primary.

A small variable capacitor is connected in series with the crystal. As a result the oscillator frequency can be changed between several hundred and several thousand hertz, depending upon the operating frequency. This control permits some shift in frequency to be made if the unit is being employed as a miniwatt QRP transmitter. Small compression-trimmer capacitors can be used for frequency changing and tuning.

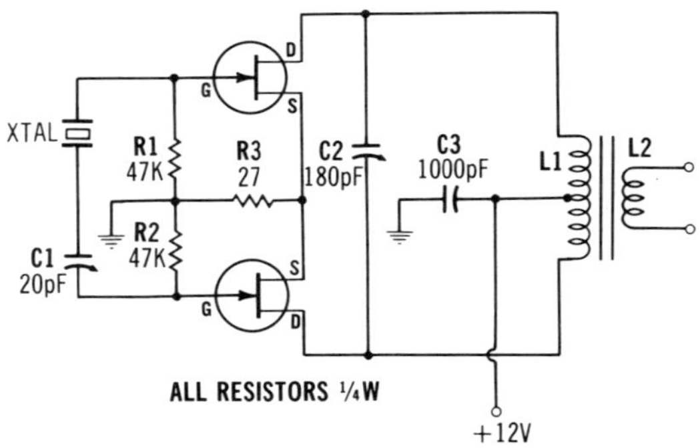

Fig. 2-14. Push-pull crystal oscillator.

FET ASTABLE MULTIVIBRATOR

A multivibrator oscillator is a two-transistor configuration that generates a square-wave or a pulse output. It is in effect a two-stage amplifier fed back on itself, with the drain of each stage linked back to the gate of the other stage (Fig. 2-15). The feedback is strong and the two FETs are alternately driven beyond cutoff. If *Q2* is initially conducting, a ground potential is placed on the top of *C1*. Capacitor *C1* thus discharges through *R1*, eventually turning on *Q1*. The top of *C2* is now at ground and thus cuts off *Q2* until *C2* has discharged sufficiently to allow *Q2* to conduct. Then the cycle repeats. If the two time constants *R1C1* and *R2C2* are of equal value and if resistors *R3* and *R4* are of proper value, a good square wave is developed at the output. This square wave can be removed at either drain of the multivibrator.

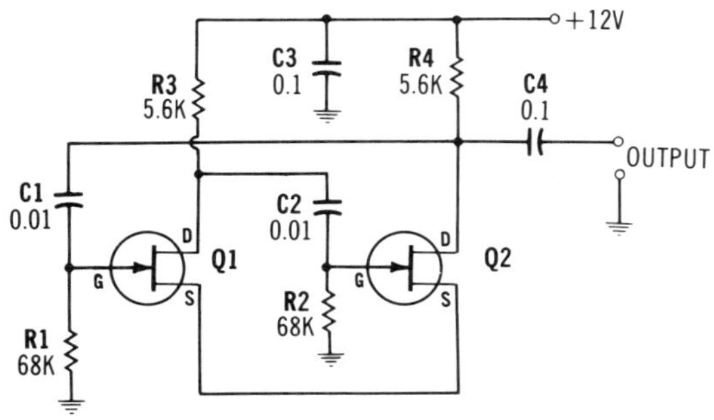

ALL RESISTORS ¼W

Fig. 2-15. FET multivibrator.

If an oscilloscope is available, the square wave can be observed. For the constants shown, the operating frequency falls between 800 and 1000 hertz. If the square-wave frequency is to be increased, it is only necessary to lower the two time constants, *R1C1* and *R2C2*, a like amount. A lower frequency is obtained by increasing these time constants.

An asymmetrical or pulselike wave can be obtained by using unequal time constants. For example, a positive pulse can be obtained by decreasing the *R2C2* time constant relative to the *R1C1* value. This will make certain that FET *Q2* is cut off for a shorter period than the first one. It is during this interval that the voltage is positive on the drain of transistor *Q2*, and therefore a positive pulse appears at the output. If desired, a negative pulse can be derived at the same time from the drain circuit of the first transistor.

If the frequency is to remain the same but a pulse output is to be obtained, it is necessary to increase the time constant *R1C1* by the same amount that the time constant of *R2C2* is decreased. The duration of the positive pulse in the output depends on how much shorter time constant *R2C2* is than time constant *R1C1*.

A multivibrator of this type is a simple circuit for obtaining a square wave or pulse output. Furthermore, the frequency or duration of the square wave or pulse can be changed conveniently. Any general-purpose field-effect transistor will function well in the circuit.

FILE 2-13

FET VFO OSCILLATOR

A stable variable-frequency oscillator for operation in the 80-meter band is shown in Fig. 2-16. Good stability begins with the use of a gate resonant circuit that has a high *C*-to-*L* ratio. Polystyrene capacitors are recommended for *C1, C2,* and *C3*. Tuning capacitor *C4* is a small broadcast variable type and, with proper adjustment of inductor *L1,* covers the frequency range from 3.5 to 4 megahertz as the capacitor is varied between maximum and minimum capacitance. Inductor *L1* can be a J. W. Miller 218336RBI or equivalent. To reduce instability and hand-capacitance effect the variable capacitor must be mounted rigidly. An insulated coupling is used between the shaft of the capacitor and any associated vernier dial.

The oscillator has high-level and low-level outputs that can be selected according to application. The load attached to the low-level output has no, or only minimal, influence on the frequency of the oscillator. Although the high output is stronger the device to which the oscillator supplies signal will have some influence on the frequency of the oscillator. This, however, can be accommodated by proper tuning of the oscillator, provided the load placed on the oscillator is of constant value. The output level and loading can be adjusted with the selection of a proper value for capacitor *C6*.

Capacitor *C7* has an influence on the low output level. If the low output is to be used for local oscillator application, the value of capacitor *C7* can be adjusted to attain optimum local-oscillator injection. If a very minimum of loading is required, a source-follower stage is advisable. It can be driven from the high output of the oscillator.

The field-effect transistor is a 2N4416. A 2N-3823, Radio Shack 276-2036, or other radio-frequency amplifier FET can be used.

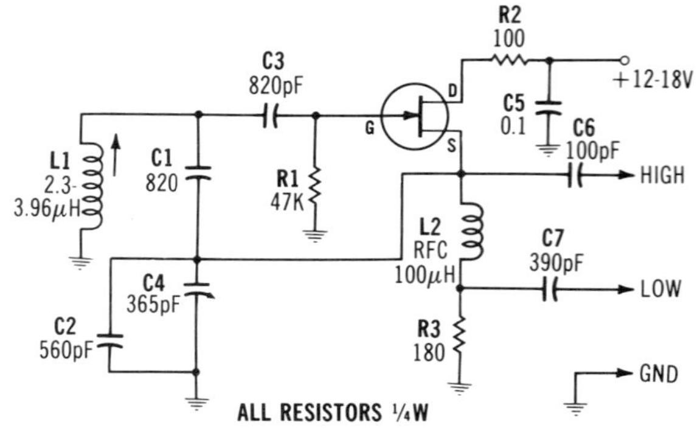

Fig. 2-16. Variable-frequency oscillator.

DIRECT-CONVERSION DETECTOR

In the direct-conversion process the incoming signal is heterodyned with a local-oscillator frequency to make a direct conversion between signal frequency and the audio modulation on the incoming signal. This type of detection is particularly successful in the demodulation of single-sideband (ssb) and continuous wave (cw) code signal.

In the direct-conversion detector of Fig. 2-17 the input resonant circuit is tuned to the 80-meter band. The input transformer assembly, *L1*, provides a match between a low-impedance antenna, such as a dipole, and the high-impedance input of the gate circuit. A small trimmer or variable capacitor can be used to tune the input resonant circuit. Inductor *L1* has an adjustable core, and this, when preset properly, permits capacitor *C1* to tune over the 3.5- to 4-MHz frequency range.

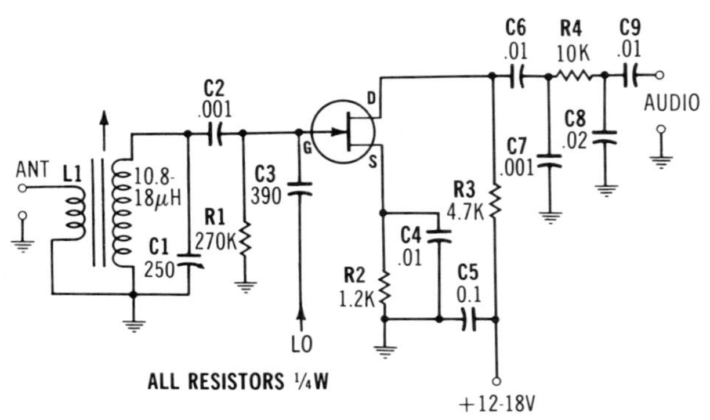

Fig. 2-17. Direct-conversion detector.

The primary winding of the input assembly consists of three turns of No. 24 wire mounted directly below the coil of the standard coil assembly (J. W. Miller 21A155RBI). The local-oscillator injection signal can be obtained from the low output of the vfo given in File 2-13.

In the drain output circuit of the detector it is the difference frequency that is emphasized. Since the incoming signal and vfo frequency are the same, there is a direct demodulation to the audio frequency. There follows an audio filter composed of resistor *R4* along with capacitors *C7* and *C8*. This filter permits only the audio-frequency components to be transferred to the follow-up high-gain audio amplifier. The original radio-frequency components and higher-order modulation components are removed by the filter.

The direct-conversion detector used in the author's example was a 2N4416. A 2N3823, a Radio Shack 276-2036, and other radio-frequency FETs can be used in the circuit.

FILE 2-15

DOUBLE-SIDEBAND GENERATOR

The double-sideband generator (Fig. 2-18) is basically an amplitude modulator which suppresses the carrier in the modulation process. The dsb generator of Fig. 2-18 consists of a push-pull crystal oscillator to serve as the carrier source and a following balanced modulator. Both stages employ the Siliconix 2N5912 dual field-effect transistor. In this device there are two transistors with identical characteristics within the same case, which is ideal for use in balanced modulators, balanced mixers, and other push-pull circuit configurations.

In the operation of the balanced modulator the phasing of the carrier input signal is such that it will be cancelled in the output. For example, the push-pull carrier component from the crystal oscillator is applied to opposite gates of the dual FET. The output drains are connected in parallel and therefore the carrier itself is cancelled. The audio signal, too, is applied in balanced fashion to the opposite sources of the dual FET.

The carrier and modulating input waves are balanced signals. In the case of the carrier this is accomplished with the use of the push-pull oscillator. Audio can be derived from a balanced modulation transformer or another source that provides equal-amplitude and opposite-polarity audio components to the input.

In the oscillator circuit the crystal is connected from gate to gate. The radio-frequency ground is located at the center tap of the coil of the drain resonant circuit. Thus equal-amplitude but opposite-polarity carriers can be derived and transferred through capacitors *C2* and *C3* to the balanced-modulator stage. Capacitor *C2* is an adjustable one and is set for best carrier suppression while the carrier component is being observed at the balanced modulator output.

Note in the balanced-modulator circuit that the two drains are connected together. The supply voltage is shunt fed through the radio-frequency choke. The output resonant circuit is single ended and uses a few-turns secondary to match a low-impedance load or antenna. A crystal diode and resistor-capacitor combination in conjunction with a 0- to 50-μA microammeter serve as an output indicator. Capacitors *C4* and *C5* are radio-frequency filters and should be mounted as near as possible to the tap on inductor *L1* and the supply voltage side of the radio-frequency choke.

Coil data can be obtained by referring to File 1-10. The crystal oscillator coil must be center

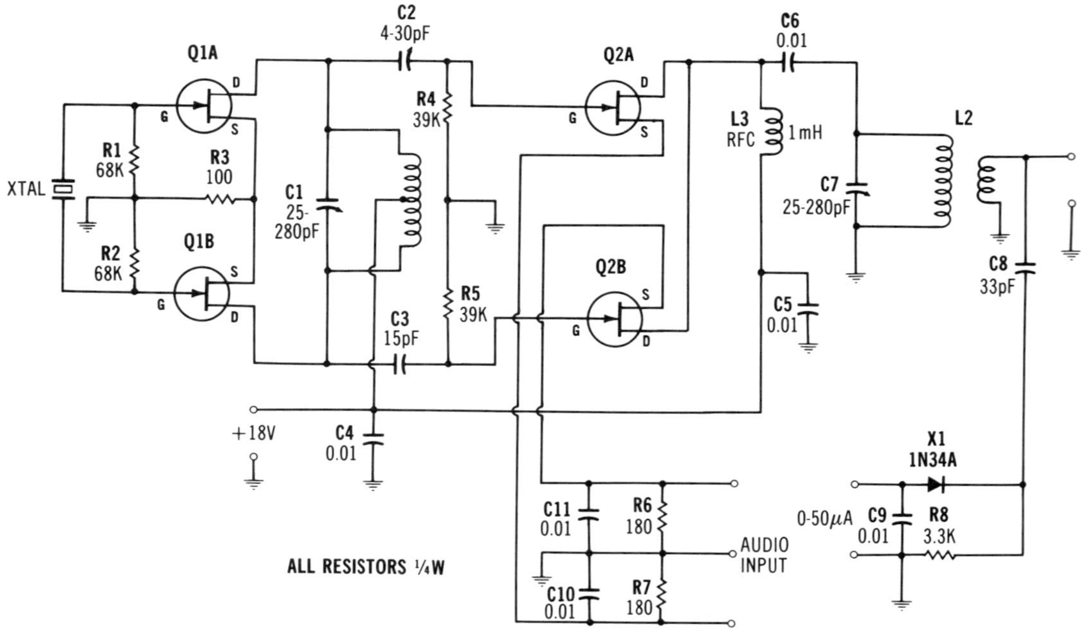

Fig. 2-18. Double-sideband detector.

tapped while a few-turns secondary must be wound around the balanced-modulator output inductor, *L2*. In the author's test circuit for 80-meter operation two toroid cores were wound on 13/16-inch (2.06-cm) cores (Permacor 57-1541). Inductor *L1* consisted of 57 turns of No. 24 enameled copper wire close wound and center tapped at the 29th turn. Inductor *L2* consisted of 56 turns of No. 24 enameled copper wire. A 4-turn low-impedance winding was wound at the center position of primary winding *L2*. Single instead of dual FETs can be made to operate in the circuit provided you select them for good, balanced operation.

FILE 2-16

VHF OSCILLATOR-AMPLIFIER

The Siliconix dual-FET 2N5912s operate well in the vhf range of frequencies. A push-pull crystal oscillator and follow-up push-pull amplifier are shown in Fig. 2-19. The crystal oscillator uses an overtone crystal that can be selected to produce an output frequency in the 6-meter radio-amateur band. The following amplifier builds up the level of this signal and supplies it to a low-impedance load or antenna system. The output stage can also be made to function as a tripler with the choice of an appropriate drain resonant circuit.

Capacitors *C1* and *C2* are 8-pF butterfly capacitors (E. F. Johnson 160-208). Inductors *L1* and *L2* consist of 7 turns of Barker and Williamson 3011 miniductor. Both coils are center tapped. Inductor *L3* consists of 2 turns of hookup wire wound around the center of inductor *L2*. Inductors *L1* and *L2* can be homemade using 7 turns of No. 22 enameled copper wire. The winding diameter is ¾ inch while the winding length is ½ inch.

Other types of dual FETs can be used in the circuit as well as four single FETs if they are selected carefully for identical characteristics.

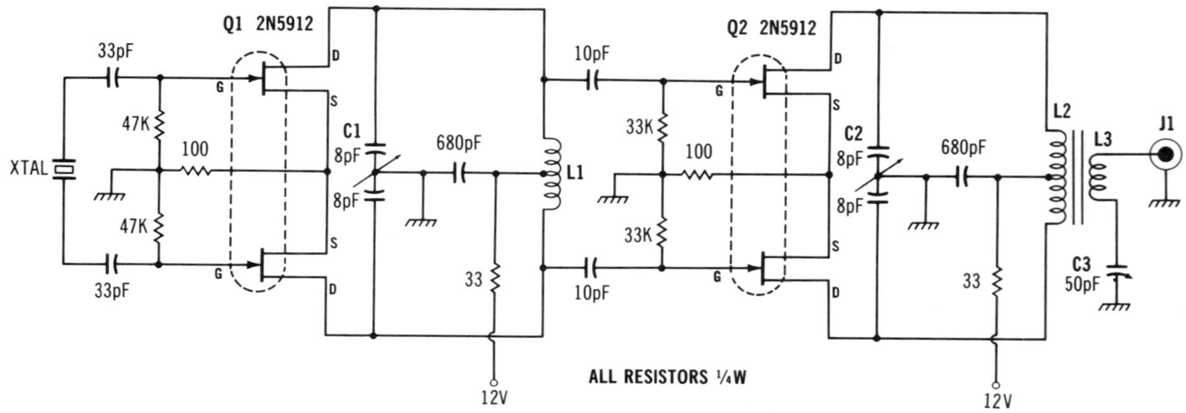

Fig. 2-19. Vhf oscillator-amplifier.

PULSE/SAWTOOTH GENERATOR

A simple pulse/sawtooth generator is shown in Fig. 2-20. The sawtooth signal can be removed at *Q2* drain, the pulse signal at the common-source output. The circuit can be operated free-running (astable) or it can be synchronized with the application of a negative spike to the gate of *Q1*.

In the generator circuit, feedback action between the two devices permits first one FET to conduct and then the other. When the second FET is not conducting, capacitor C3 charges through resistors *R6* and *R7*, generating the sawtooth trace as shown in Fig. 2-21. When the second FET conducts, the capacitor discharges rapidly through the low resistance presented by the conducting device, forming the fast sawtooth retrace. The return, or retrace, portion of the waveform is much faster because the capacitor discharges through a low resistance whereas it charges through the high resistance contributed by resistors *R6* and *R7*. The amplitude of the sawtooth is controlled with potentiometer *R6*.

Capacitor *C2* and resistors *R3* and *R4* serve as the frequency-determining resistor-capacitor combination. The time that *Q2* remains cut off is determined by the time constant of these three components. Resistor *R4* is a potentiometer that can be used to control the frequency of the pulse/sawtooth generator.

The waveforms of Fig. 2-21 demonstrate the circuit activity. The waveform at the gate of *Q2* shows the long nonconducting period of this FET. During this interval the sawtooth-forming capacitor is charging slowly. When the conduction level is reached, the gate is driven positive very quickly because of the feedback path between the two FETs.

Two Siliconix 2N4339s were used in the author's setup. Other small-signal transistors can be used in the circuit.

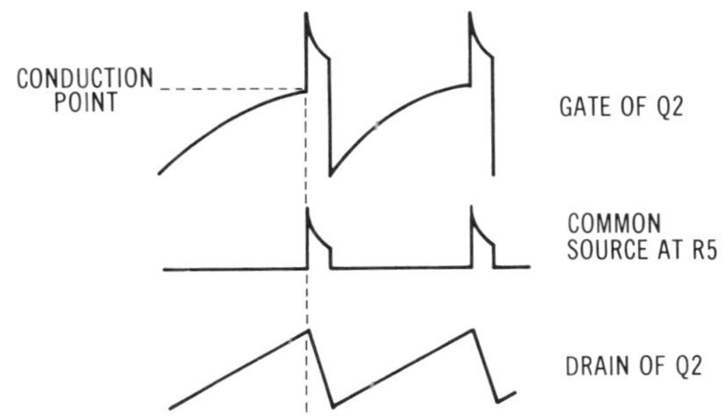

Fig. 2-21. Typical waveforms.

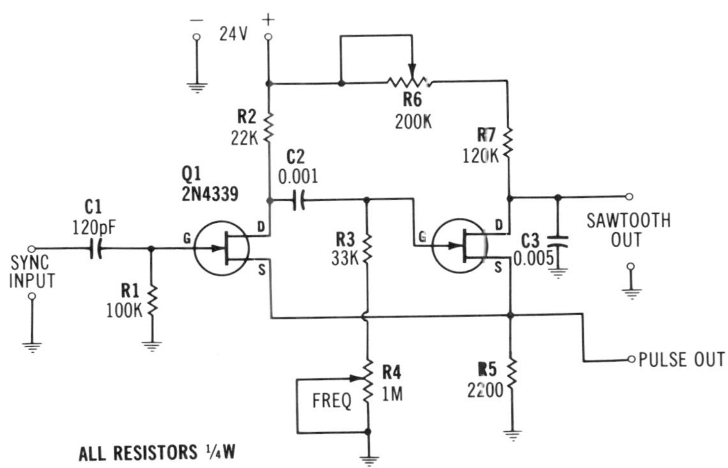

Fig. 2-20. Pulse/sawtooth generator.

FILE 2-18

MOSFET COMMON-SOURCE AMPLIFIER

The structure of the MOSFET (metal-oxide semiconductor field-effect transistor) provides some unique characteristics and makes the MOSFET attractive for small-signal and low-noise radio-frequency amplifiers and mixers. The characteristics are favorable for use in other high-impedance applications and in chopper and switching circuits. In the structural arrangement (Fig. 2-22A) the gate electrode acts as a control element, just as it does in a junction field-effect transistor. However, that control is exerted whether the gate is swinging positive or negative. In the normal operation of a junction-type FET the gate exerts an influence only when the junction is reverse biased.

In the MOSFET construction a thin gate layer is insulated from the semiconductor material that

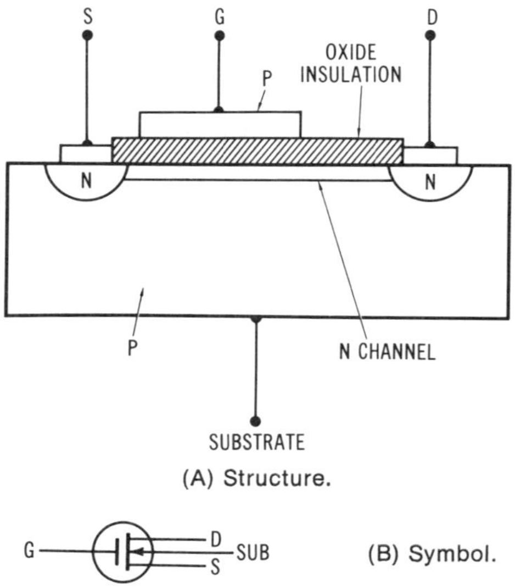

Fig. 2-22. Depletion-mode MOSFET.

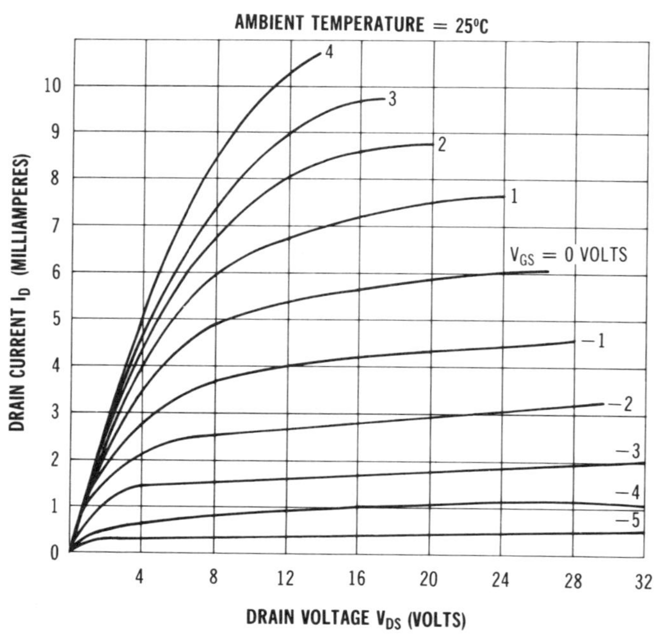

Fig. 2-23. Characteristic curves of depletion-mode MOSFET.

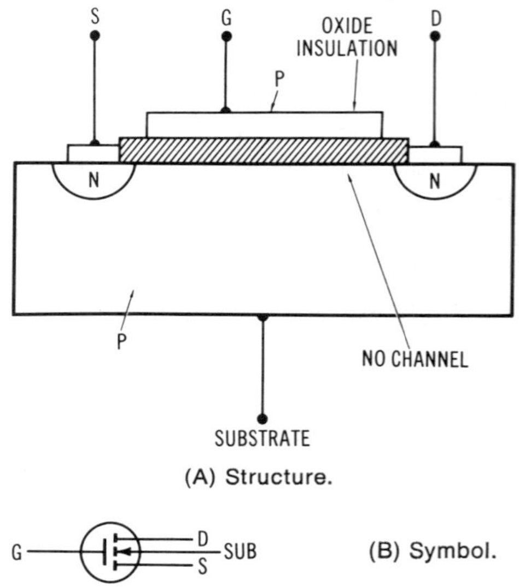

Fig. 2-24. Enhancement-mode MOSFET.

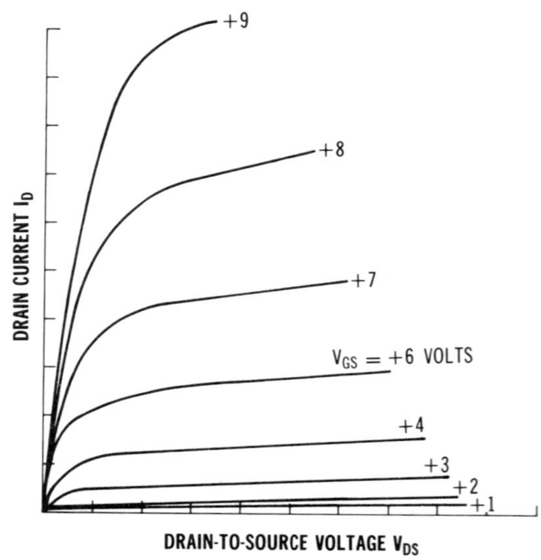

Fig. 2-25. Enhancement-mode MOSFET curves.

forms the channel by a thin layer of silicon dioxide. The electrical field that results when the gate is biased appropriately influences the active carriers in the source-drain channel. Since there is insulation instead of an actual junction between the gate and channel, the leakage current is very low and is little affected by temperature. A very high input impedance results. The gate and channel, separated by an oxide layer, are in effect the plates of a capacitor.

The arrangement of Fig. 2-22A shows the structure of a depletion-type MOSFET. In this example the channel material is n type and there are excess electrons. As the gate is made negative relative to the source the number of electrons in the channel is depleted (reduced). This action is similar to that of a junction FET. During normal operation, however, the gate of a junction FET may not be permitted to swing significantly past zero because the junction will then be forward biased, decreasing the input impedance. The MOSFET has no such limitation. If the metal gate is made to swing past zero there is no increase in input current or decrease in input impedance, and the gate maintains its control of channel conductivity. This can be seen in the characteristic curves of Fig. 2-23. Note that the gate voltage can be made to swing either positive or negative.

A more popular type of MOSFET construction is the enhancement-mode arrangement of Fig. 2-24. Note that in its basic structure there appears to be no actual channel between the source and the drain. However, the gate itself does span this region and when a positive bias is applied to the gate, a conduction path between the source and the drain is established. The conductivity of this path is a function of the charge on the gate. In this manner the magnitude of the drain current is controlled.

The unusual characteristic of the enhancement-type MOSFET is that it is normally cut off (no drain current) until the application of the positive gate voltage. The application of a gate voltage causes the drain-source current to increase. The more positive the gate-source voltage, the higher is the drain current. This is shown in the characteristic curves of Fig. 2-25.

The high input impedance is both an advantage and a hazard of the MOSFET device. Since the input impedance is so high, a very tiny current or static charge is able to build up a high voltage on the gate, which can possibly damage the device. Thus MOSFETs must be handled carefully and circuit arrangements must be such that excessive charges are kept off the gate. Another way of avoiding this hazard is by fabricating gate-protective diodes within the device (Fig. 2-26). These diodes are electrically connected between gate and source, bypassing those voltage transients which could damage the device.

The two- or dual-gate arrangement provides a separate gate for signal and for applying an external control bias. The dual-gate configuration can be used conveniently in mixer and other balanced circuit arrangements. The two-gate plan also makes possible a reduction in feedback capacitance obtained by ac grounding of the second gate. Thus the device can be used favorably in amplifier circuits without neutralization. Local oscillator feed through to the antenna is reduced.

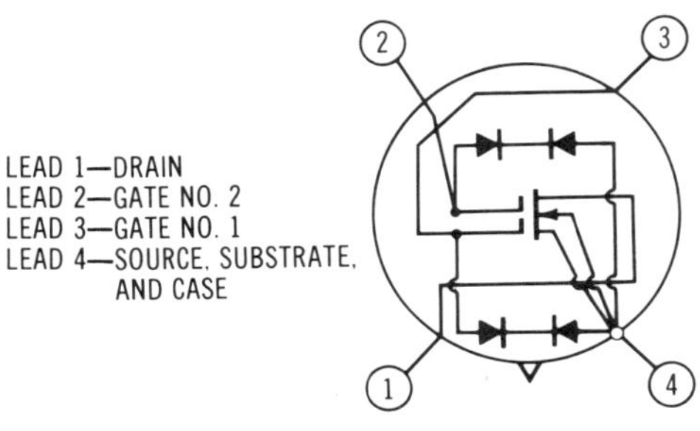

Fig. 2-26. Dual-gate MOSFET with protective diodes.

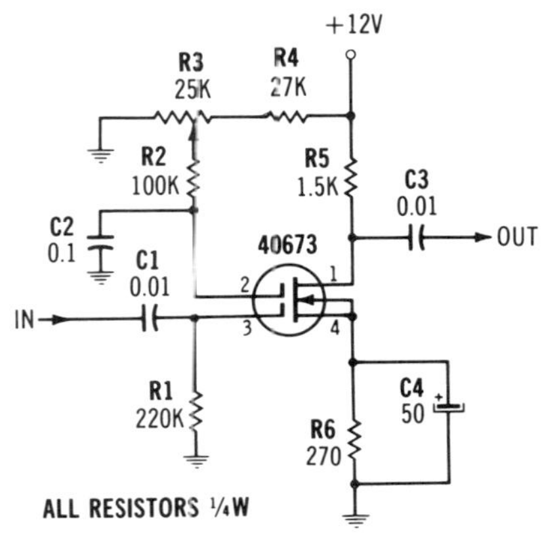

Fig. 2-27. Basic dual-gate common-source circuit.

A popular dual-gate MOSFET used widely in radio-frequency amplifiers and mixers is the 40673. A test circuit is given in Fig. 2-27. The basic characteristics can be checked with an audio-frequency signal. Instead of a resonant circuit, resistor *R5* serves as the drain-circuit load.

The signal is applied to gate No. 1 (pin 3). The gain of the amplifier can be regulated by applying a dc control voltage to gate No. 2 (pin 2). The input and output waves can be observed on an oscilloscope. In this circuit arrangement the gain will not be high but sufficient to demonstrate operation.

Set potentiometer *R3* to midposition. Display the output wave on the oscilloscope screen. Now vary the potentiometer. Note the influence that the control voltage on the second gate has on the gain of the device.

VMOS FET POWER OSCILLATOR

The conventional JFET and MOSFET transistors are fundamentally low-power devices. The regular FET is a horizontal-motion device (Fig. 2-28) because the charge motion along the body or channel flows horizontally from source to drain. This limits the current density, compelling the chip area to be large and the cost high if any significant power capability is to be obtained.

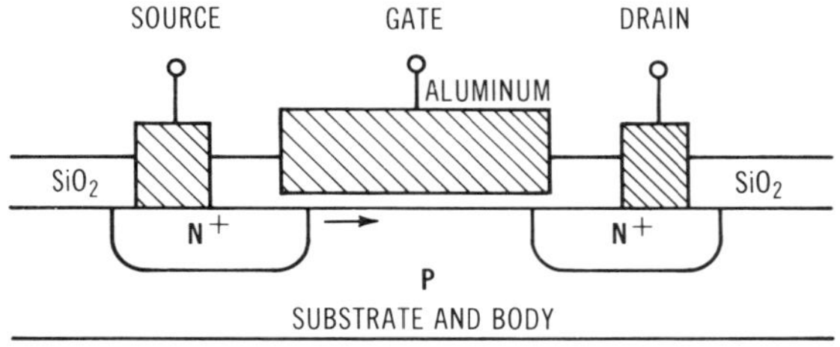

Fig. 2-28. Cross section of a conventional MOSFET.

This power limitation, however, has been eliminated with the development of the VMOS (vertical metal-oxide semiconductor) field-effect transistor (Fig. 2-29). The V-groove gate region and associated arrangement of elements permits a vertical motion of charges, and a much greater current density is possible. It is an enhancement-mode transistor and the gate and drain must both be positive relative to the source if there is to be a motion of charges. Note the characteristics of Fig. 2-30. Before there is drain current the gate voltage must be made positive. At 0 volts and negative voltages the drain current is cut off.

In operation the positive gate sets up an n-type channel on both sides of the p-type body. Electrons flow from the source through the n-type channel and the EPI (epitaxial) region to the substrate. The substrate serves as the drain connection for the VMOS device. The higher width-to-length ratio of the VMOS channel plus other factors provide a much greater current density and a higher power-handling capability. The power capability is now comparable to that of bipolar devices without some of their disadvantages.

The VMOS is a high-impedance voltage-drive device that provides good linearity and low distortion. VMOS devices function well as either weak-

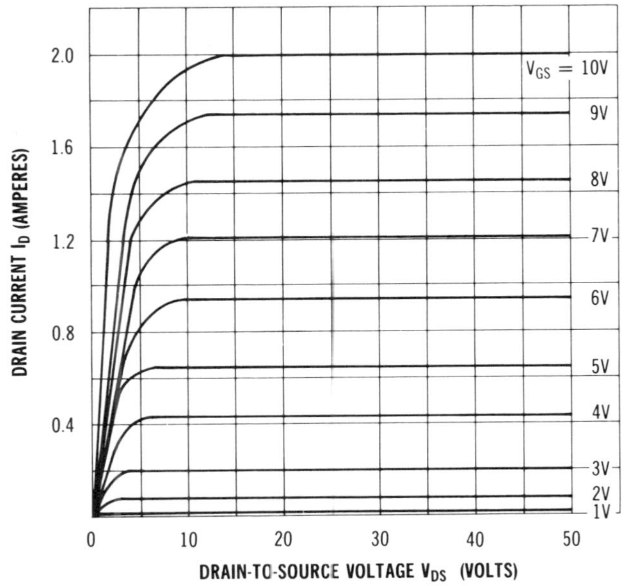

Fig. 2-30. VMOS characteristic curves.

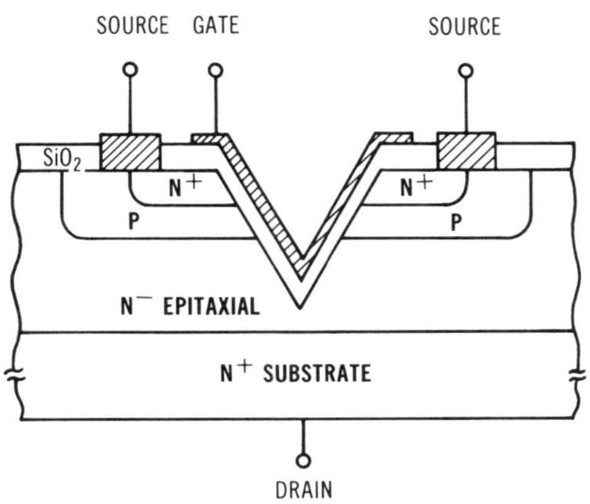

Fig. 2-29. Cross section of a VMOS channel.

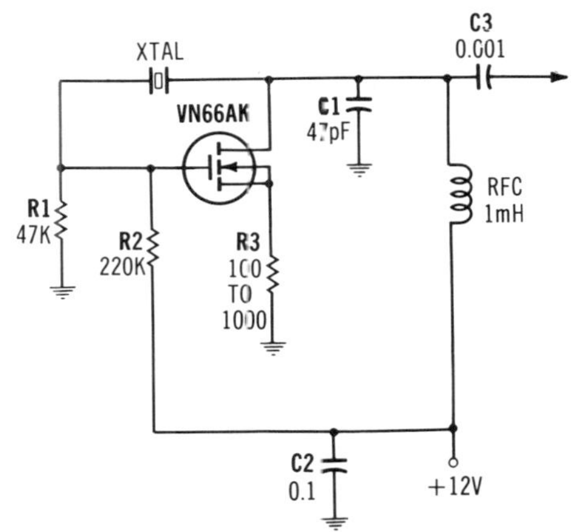

Fig. 2-31. VMOS crystal oscillator.

or strong-signal amplifiers. As the device heats up its characteristic is to draw less current instead of more. Consequently, the thermal-runaway hazard does not exist. It is not subject to secondary or mismatch breakdown. Furthermore, there is no change in input capacitance with changes in drive level and frequency.

A VMOS connected in a Pierce crystal oscillator circuit is shown in Fig. 2-31. The VMOS can be a Siliconix VN66AK or Supertex VN2 device. A good output is obtained using a 12-volt supply. A higher output can be obtained by increasing the drain voltage to 18–24 volts. The circuit differs from the conventional FET arrangement because the VMOS is an enhancement-mode device and it is necessary to apply a starting bias by way of resistors *R1* and *R2*. The value of resistor *R3* can be regulated to control the output level and shape of the output wave. In general, the lower the output, the more sinusoidal is the output waveform. The oscillator will operate on fundamental crystals from 1 to 25 MHz and possibly higher with adjustment of component values.

It is interesting that the very same circuit will function when substituting a conventional FET (Radio Shack 276-2035) or a bipolar transistor (2N2222). Many other devices will function well in this circuit.

VMOS TUNED-OUTPUT OSCILLATOR

If a low-impedance output is to be obtained using a crystal oscillator circuit, the arrangement of Fig. 2-32 is a good choice. A good low-impedance drive is obtained. Resistors *R1* and *R2* provide starting bias for the Siliconix VMOS device. Resistor *R3* can be used to adjust the output level.

The *L1-C2* combination should be resonant at the crystal frequency. Coil data can be obtained from File 1-10. In the author's circuit for 40-meter operation 35 turns of No. 26 enameled wire were closewound on an Amidon T-50-2 toroid (red). The secondary consisted of 4 turns of No. 26 mounted over the center of the primary. This arrangement provided a suitable match to a 50-ohm termination. The output indicator was the same as that employed in File 1-9. The output wave can be observed on an oscilloscope.

The oscillator is a versatile circuit, and other VMOS, JFET, and bipolar transistors can be used in the circuit. Some adjustment of the value of resistor *R3* may be necessary to obtain easy starting and the desired output level.

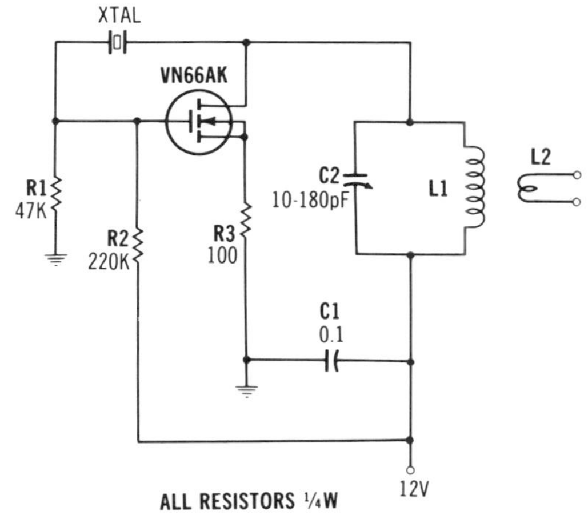

Fig. 2-32. VMOS oscillator with tuned output.

FILE 2-21

VMOS OSCILLATOR-AMPLIFIER

A VMOS device operates well as an rf power amplifier. Such an amplifier arrangement is shown in Fig. 2-33. Included is the Pierce crystal oscillator, which provides adequate voltage drive for the VN66AK (Siliconix) rf amplifier stage. Coil data for the rf output stage can be obtained from File 1-10. Note that the primary is center tapped to provide a lower-impedance output for the VMOS power device. In the author's example, for 40-meter operation, inductor *L1* was a T-50-2 toroid (red) wound with 35 turns of No. 26 enameled copper wire. This primary is center tapped. The secondary consists of 4 turns of No. 26 wire wound around the center of the coil.

The output can be used to drive the File 1-9 indicator. A 47-ohm resistor is used as a termination for the secondary winding. An oscilloscope can be used to observe the output waveform.

The VMOS transistor is an enhancement-mode device and therefore adequate drive must be obtained from the crystal oscillator. The positive alternation at the gate drives the output transistor into conduction. Resonant voltage is developed across the output resonant circuit by the bursts of drain current.

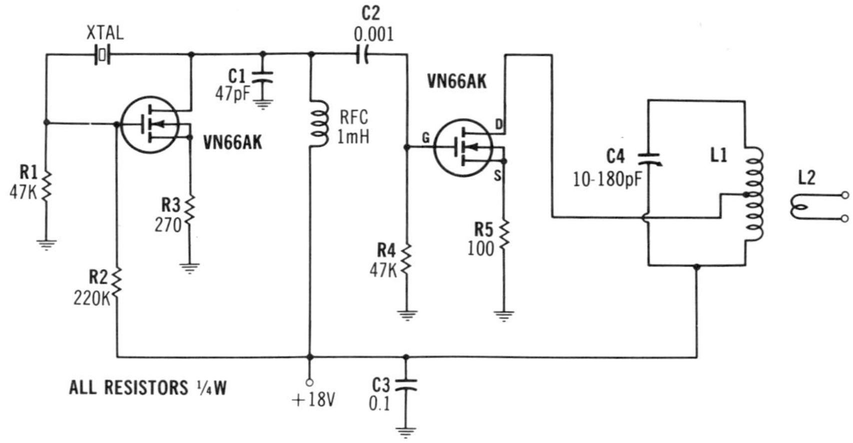

Fig. 2-33. VMOS oscillator-amplifier.

FILE 3

LINEAR INTEGRATED CIRCUITS

AUDIO POWER AMPLIFIER

An integrated circuit (IC) consists of from a few to hundreds of individual solid-state devices packaged in a single container. The major components are transistors, diodes, and resistors in variout configurations forming simple or complex electronic systems. An example of an audio power amplifier IC is shown in Fig. 3-1.

The LM386 is basically a 1-watt audio power amplifier that begins with a differential input stage consisting of Darlington input pairs that raise the input resistance. Note that the collector output of the Darlington pair associated with pin 3 is connected to the base circuit of the push-pull output stage. The inputs are ground referenced, the signal usually being applied to input 3 while input 2 is grounded. The output is biased automatically to one-half of the supply voltage. The quiescent or no-signal power drain of the LM386 is only 18 milliwatts when operating from a 6-volt supply. The power demand, of course, rises with the level of the input signal.

The fundamental gain of the LM386 is 20 with the simple circuit arrangement of Fig. 3-1C. Additional gain can be obtained by connecting external components between pins 1 and 8. Note that these components are, in effect, shunted across the internal 1.35-kilohm resistor, which is in the emitter-current paths of the input transistors. As shown in Fig. 3-1D the gain of the amplifier can be increased to 200 by shunting a 10-μF capacitor between pins 1 and 8. If a resistor is placed in series with a capacitor between pins 1 and 8, the amplifier gain can be set to any value between 20 and 200.

The File 3-1 practical circuit of Fig. 3-2 provides a voltage gain of about 50. The supply voltage can be between 4 and 12 volts. A greater power output is obtained at the higher voltages. Poten-

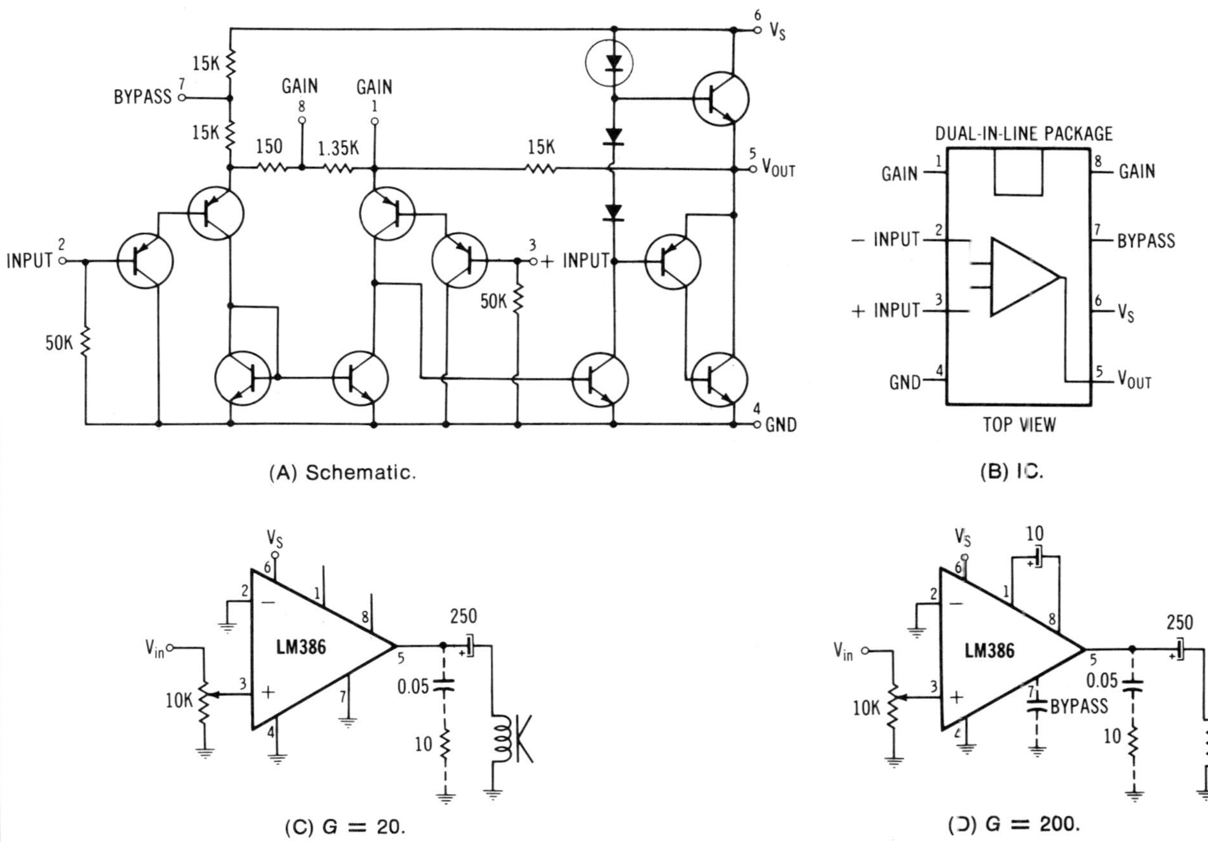

Fig. 3-1. LM386 1-watt audio power amplifier.

tiometer *R1* permits control of the input signal level. A gain of approximately 20 can be obtained by disconnecting resistor *R2* and capacitor *C3* from between pins 1 and 8. Some boost in low-frequency signal as applied to the speakers can be obtained by increasing the value of capacitor *C6*.

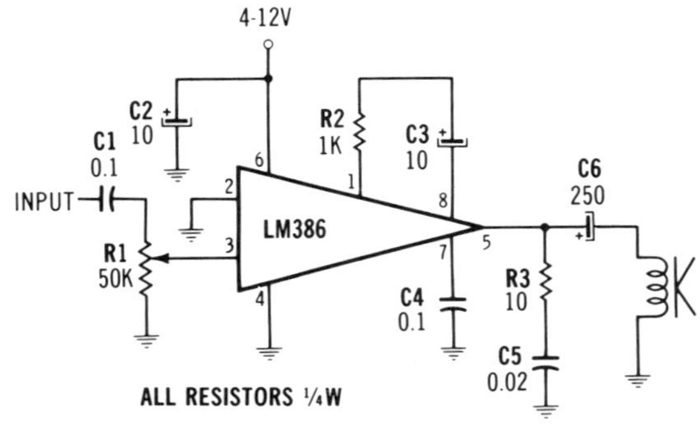

Fig. 3-2. IC audio power amplifier.

You may wish to keep the circuit of Fig. 3-2 built up permanently for use in some of the linear IC file circuits that follow. An essential piece of test equipment for working on linear and digital integrated circuits is the oscilloscope. A dual-trace type, such as the Sencore model shown in Fig. 3-3, is particularly helpful because it can be used to observe input and output waveforms simultaneously. It is particularly helpful in making comparisons between pulse and other nonsinusoidal waveforms. The dual-trace capability along with an accurately calibrated and triggered sweep permits accurate measurements of the various waveforms developed by integrated circuits.

Such a dual-trace circuit can be connected conveniently to the circuit of Fig. 3-2 to observe input and output sine waves. As the input wave is held constant in amplitude the linear or nonlinear frequency characteristic of the amplifier (depending on the circuit configuration used) can be observed. Also the input waveform can be adjusted finely to

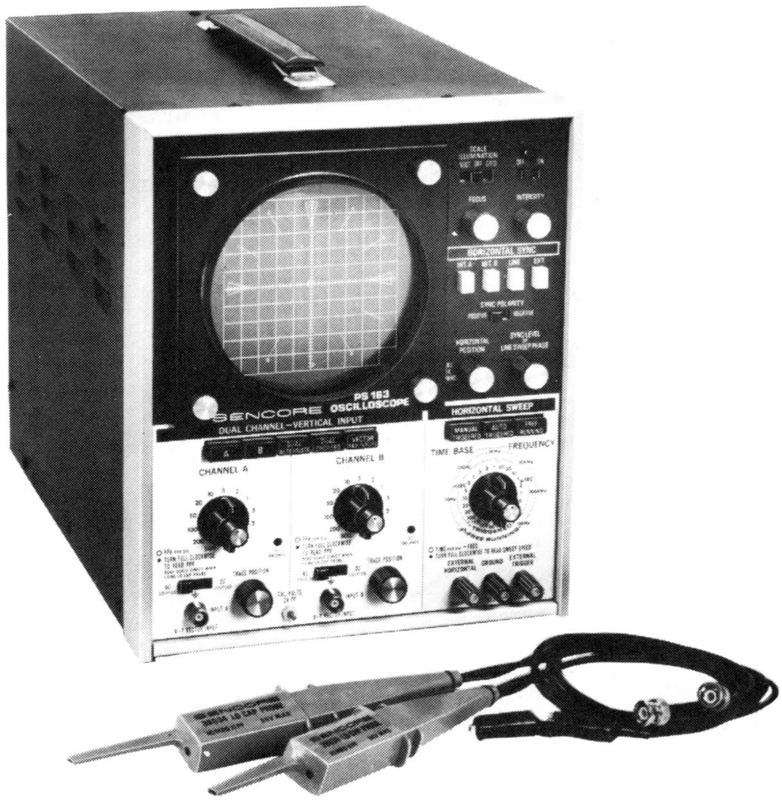

Fig. 3-3. Dual-trace oscilloscope.

Courtesy Sencore, Inc.

obtain maximum nondistorted power output. Thus measurements can be observed at the same time that maximum undistorted signal is being developed across the output.

The circuit of Fig. 3-4 shows how the amplifier can be connected as a base-boost system. In this circuit a series resistor-capacitor feedback combination is connected between pins 5 and 1. In the arrangement shown there is only a limited amount of feedback at low frequencies, while the feedback is significantly higher over the middle- and high-frequency audio range. Thus the lows are unattenuated and their output level rises above the middle and high frequencies. Consequently the circuit arrangement functions as a low-frequency boost system. The maximum gain is concentrated in the 100- to 300-hertz frequency range with low-frequency output available as low as 40 hertz.

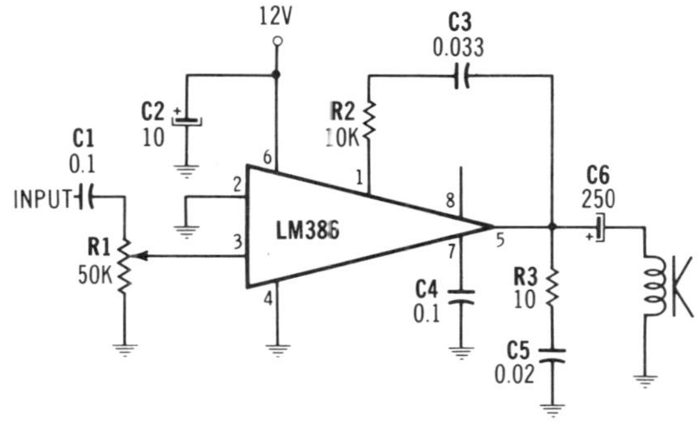

Fig. 3-4. Bass-boost circuit.

FILE 3-2

OPERATIONAL AMPLIFIERS

The operational amplifier is a dc and/or ac amplifier of high stability. It can operate with low gain or high gain, depending on the values of three external resistors. In fact, the actual gain of the amplifier depends on external resistance ratios rather than internal components.

The two common arrangements are shown in Fig. 3-5. Fig. 3-5A shows a noninverting amplifier, with the amplified output voltage being of the same polarity as the input. Balanced operation occurs when the input resistances at the input terminals are identical. In the calculation of the input resistance it is always necessary to consider the influence of the feedback resistance as shown in the product-over-sum relations. In the case of the noninverting amplifier, the input 1 resistor is of value *R1*. However, the signal input resistor should have an ohmic value that matches the parallel combination of resistors *R1* and *R2*.

Typical gain and resistance values are listed in the chart. Note that gain increases when there is a higher ratio between the ohmic values of resistor *R2* and resistor *R1*. This is to be anticipated because there is less feedback. Likewise, the bandwidth is reduced with a higher resistance ratio.

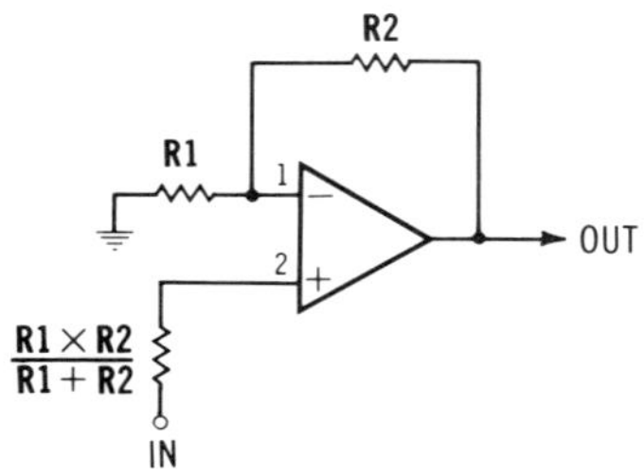

GAIN	R1	R2	BW
10	1K	10K	100kHz
100	100	10K	10kHz

(A) Noninverting mode.

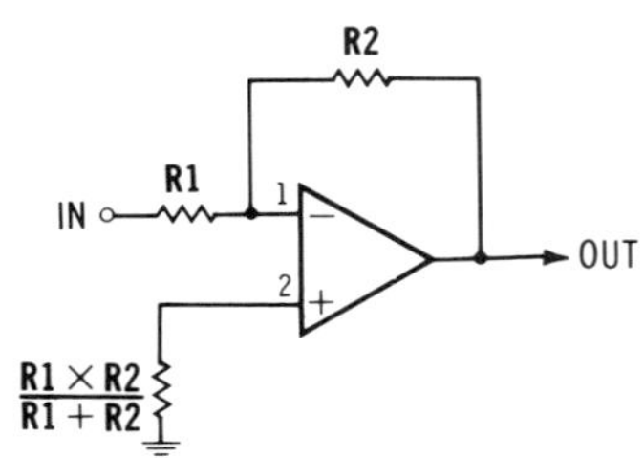

GAIN	R1	R2	BW
10	1K	10K	100kHz
100	1K	100K	10kHz

(B) Inverting mode.

Fig. 3-5. Operating modes of operational amplifier.

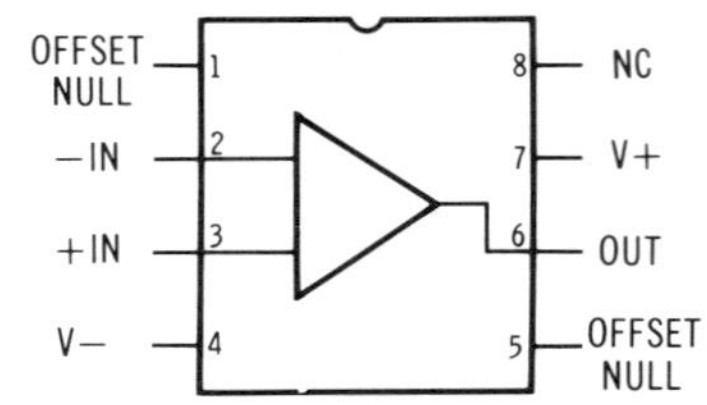

Fig. 3-6. Diagram of 8-pin minidip 741.

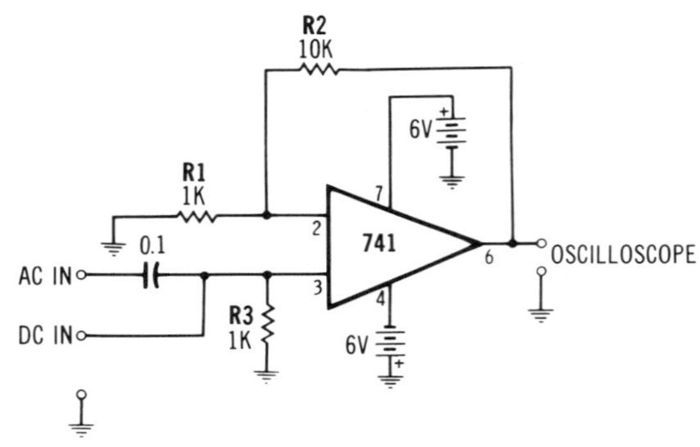

(A) Noninverting.

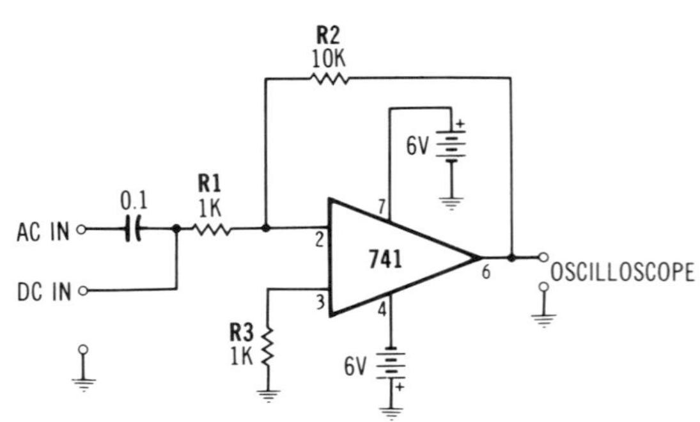

(B) Inverting.

Fig. 3-7. Circuits for operational amplifier.

An inverted output is obtained with the circuit of Fig. 3-5B. Again the resistance values determine the stage gain. Stage gain approximates the ratio of the feedback resistance *R2* over the input resistance *R1*. This can be verified from the chart by dividing the ohmic value of resistor *R1* into that of resistor *R2*.

A popular operational amplifier is the 8-pin minidip 741 (Fig. 3-6). Practical circuits for noninverting and inverting operation are given in Fig. 3-7. Two inputs are indicated, showing that for strictly ac operation an input capacitor is needed. For combination ac/dc or strictly dc operation the input signal is direct coupled. A split-voltage supply is used, e.g. two 6-volt batteries.

The two circuits of Fig. 3-7 have gains of approximately 10. Higher gains can be obtained using the values shown in the chart of Fig. 3-5 for both the noninverting and inverting modes. A dual-trace oscilloscope connected to input and output can be used to measure gain and to check the frequency response. A low dc voltage can also be applied to the inputs to check the dc amplifying characteristics of the operational amplifier. The connection of 0.5 volt dc should result in about a 5-volt dc voltage change at the output, viz, +5 volts for the circuit of Fig. 3-7A and −5 volts for the circuit of Fig. 3-7B.

FILE 3-3

SINGLE-SUPPLY OPERATIONAL AMPLIFIER

The File 3-2 circuits show the classic way of connecting operational amplifiers using a split power supply. File 3-3 shows how a single 12-volt power supply can be used to supply dc voltage to the 741 operational amplifier. A voltage divider in the form of resistors *R2* and *R3* is arranged at the input 3 terminal of the operational amplifier. This establishes proper biasing when 12 volts is connected to pin 7 and the common connection is made to pin 4. The dc voltage at the output pin 6 will now be 6 volts positive instead of 0 dc volts of the split-supply arrangement.

The circuit of Fig. 3-8 is connected as an inverting type with signal being applied to the pin 2 input. Two values are suggested for resistor *R4*. When the resistor value is 10 kilohms the amplifier will have an approximate gain of 10. If the feedback resistor *R4* has a value of 100 kilohms, the amplifier gain will be approximately 100. This gain factor can be checked accurately with a dual-trace oscilloscope. Especially when using the 100-kilohm resistor it is important that a very low level of signal be applied to the input so as not to overload the input circuit and produce a distorted output. However, the operational amplifier can be used as a square-wave generator by overdriving the input with a sine-wave signal. This is particularly the case when feedback resistor *R4* has a value of 100 kilohms.

The 741 stage of Fig. 3-8 can be used as a weak-signal audio amplifier. Its output can be used to supply signal to the audio power amplifier of File 3-1 to obtain a very high overall gain. The input signal, of course, must be of very low amplitude so as not to overdrive the combined amplifiers. Both operational amplifier and audio power amplifier can be operated at 6 volts. For even higher output the two stages can be operated from a supply of 12 volts.

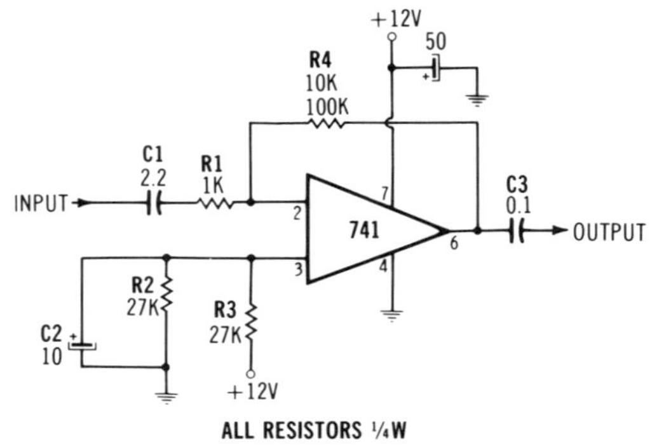

Fig. 3-8. Single-supply operational amplifier.

VOLTAGE-FOLLOWER OPERATIONAL AMPLIFIER

The operational amplifier can also be connected as a voltage follower presenting a high-impedance input and a low-impedance output. The voltage follower is similar in its characteristics to a bipolar emitter-follower or an FET source-follower. Its gain is approximately unity.

The voltage follower circuit is very simple, as shown in Fig. 3-9. A split-voltage power source is used. This can be two 6-volt lantern batteries.

In addition to the high input impedance the operational amplifier follower can accommodate a wide range of input signal levels—from very low up to several peak volts. The maximum input signal for the circuit of Fig. 3-9 is approximately 2 volts peak before any distortion of the output sine wave occurs.

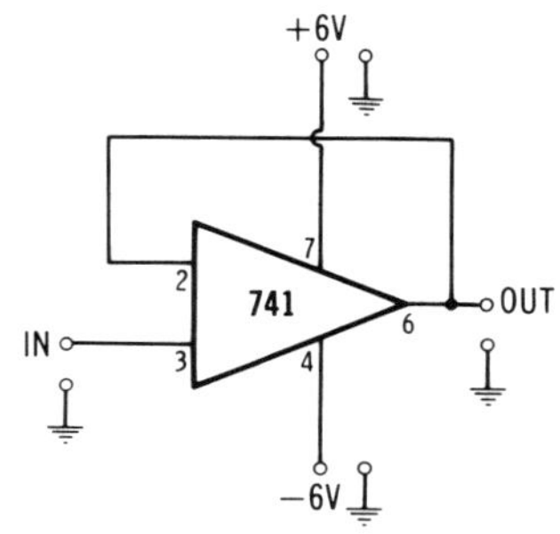

Fig. 3-9. Voltage-follower operational amplifier.

FILE 3-5

HIGH-PASS FILTER

An operational amplifier connected as a voltage follower and including an appropriate resistor-capacitor network performs well as an active filter. In a high-pass circuit (Fig. 3-10) signal frequencies are allowed to pass with minimal attenuation down to a certain frequency. Below this so-called cutoff frequency, signals are attenuated.

The cutoff frequency f_c can be determined by the equation:

$$f_c = \frac{1}{2\pi\sqrt{R1R2C1C2}} \qquad (R1 = R2,\ C1 = 2C2)$$

where *R1*, *R2*, *C1*, and *C2* are as shown in Fig. 3-10. In the circuit arrangement of Fig. 3-10 the cutoff frequency is in the 100-hertz range. This high-pass filter can be checked out conveniently with an audio generator and a dual-trace oscilloscope. Note that a split power supply is used to connect −6 volts to pin 4 and +6 volts to pin 7.

To change the cutoff frequency of the filter, multiply the ohmic value of the two resistors by the ratio of the original frequency over the new frequency:

$$R_{new} = R_{old} \times \frac{f_c\,(\text{old})}{f_c\,(\text{new})}$$

where R_{new} is the new value of *R1* and *R2*. Thus if you wish to raise the cutoff frequency to 200 hertz in the example of Fig. 3-10 it would be necessary to halve the values of resistors *R1* and *R2*. The same result can be obtained by halving the capacitor values. If both resistor and capacitor values were halved, the cutoff frequency would be increased to 400 hertz (quadrupled).

If you wish to lower the cutoff frequency to 50 hertz, you have to double the ohmic values of resistors *R1* and *R2* or double the *C1* and *C2* capacitor values.

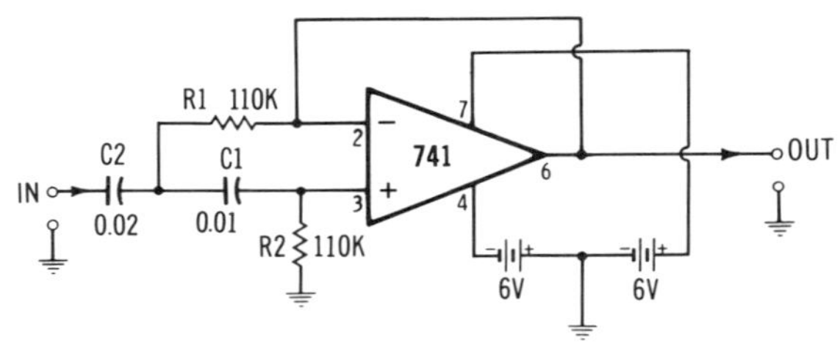

Fig. 3-10. High-pass filter.

LOW-PASS FILTER

An operational amplifier connected as a voltage follower and including an appropriate resistor-capacitor network performs well as an active filter. In a low-pass circuit (Fig. 3-11) signal frequencies are allowed to pass with minimal attenuation up to a certain frequency. Above this so-called cutoff frequency, signals are attenuated.

The cutoff frequency f_c can be determined by the equation:

$$f_c = \frac{1}{2\pi\sqrt{R1R2C1C2}} \qquad (R1 = R2,\ C1 = 2C2)$$

where *R1*, *R2*, *C1*, and *C2* are as shown in Fig. 3-11. In the circuit arrangement of Fig. 3-11 the cutoff frequency is in the 10,000-hertz range. This low-pass filter can be checked out conveniently with an audio generator and a dual-trace oscilloscope. Note that a split power supply is used to connect −6 volts to pin 4 and +6 volts to pin 7.

To change the cutoff frequency of the filter, multiply the ohmic value of the two resistors by the ratio of the original frequency over the new frequency:

$$R_{\text{new}} = R_{\text{old}} \times \frac{f\,(\text{old})}{f\,(\text{new})}$$

where R_{new} is the new value of *R1* and *R2*. Thus if you wish to raise the cutoff frequency to 20,000 hertz in the example of Fig. 3-11, it would be necessary to halve the values of resistors *R1* and *R2*. The same result can be obtained by halving the capacitor values. If both resistor and capacitor values were halved, the cutoff frequency would increase to 40,000 hertz (quadrupled).

If you wish to lower the cutoff frequency to 5000 hertz, you have to double the ohmic values of resistors *R1* and *R2* or double the *C1* and *C2* capacitor values.

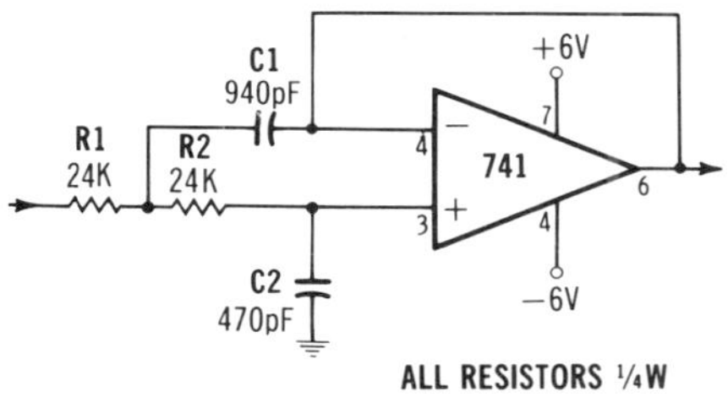

Fig. 3-11. Low-pass filter.

FILE 3-7

BANDPASS FILTER

A combination low-pass and high-pass resistor-capacitor pair can set up a bandpass amplifier as shown in Fig. 3-12. Resistors *R1* and *R2* along with capacitors *C1* and *C2* form the basic feedback arrangement. Optional potentiometer *R3* and 1-kilohm resistor *R4* provide a means of shifting the cutoff frequency over a limited range if desired.

The circuit of Fig. 3-12 has a peak response in the 600- to 800-hertz range, depending on parts tolerances. The optional potentiometer permits a more precise adjustment of the peak frequency. Bandwidth and gain are determined by the basic four components. These also determine the peak frequency. The peak frequency itself, however, can be varied with optional potentiometer *R3* and the 1-kilohm resistor. Some adjustment of *R2* may also be necessary in setting the exact peak frequency.

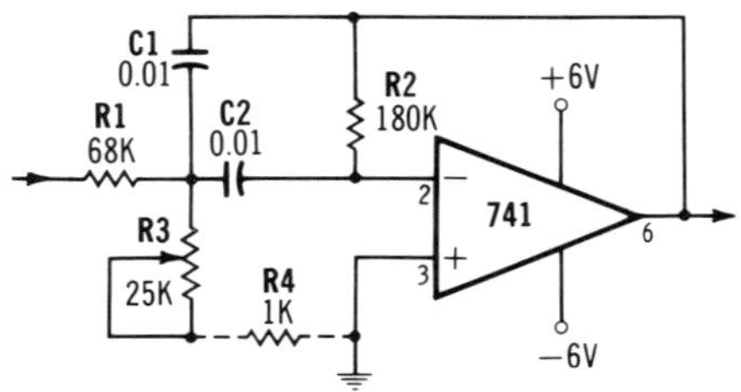

ALL RESISTORS ¼W

Fig. 3-12. Bandpass filter.

WIDEBAND BANDPASS FILTER

If a wideband filter is desired, a high-pass and a low-pass filter can be cascaded as shown in Fig. 3-13. This circuit combines the high-pass filter of File 3-5 and the low-pass filter of File 3-6. Since the high-pass filter has a cutoff of approximately 10 kilohertz and the low-pass filter a cutoff of 100 hertz, the response of the combination will be essentially flat between 100 and 10,000 hertz.

Low- and high-frequency cutoff frequencies can be adjusted by changing the appropriate values for either the high-pass or low-pass side. In two-way radio communications it is customary to use a bandwidth of approximately 300 to 3000 hertz. To obtain this passband it is necessary to reduce the high-frequency cutoff by approximately one-third and to triple the low cutoff frequency. One approximate way of doing this is to reduce the ohmic value of resistors *R1* and *R2* to one-third of their present value and to triple the values of capacitors *C3* and *C4*. A precise bandpass of course can be obtained by using the basic equation given in File 3-5 and close-tolerance parts values. Remember that for each filter the two resistors should be of equal value while one capacitor is double the value of the other.

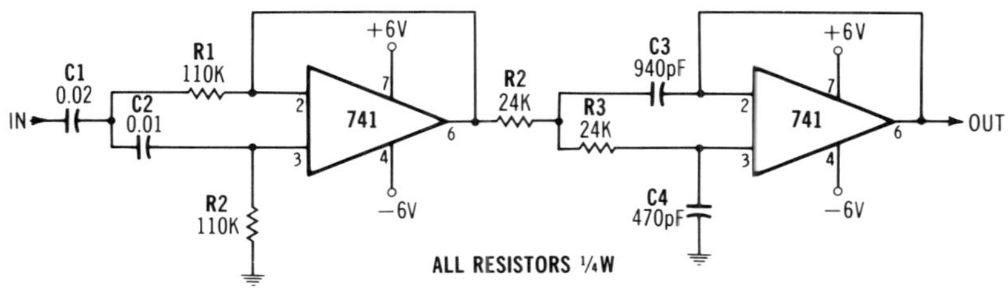

Fig. 3-13. Wideband bandpass filter.

FILE 3-9

ASTABLE PULSE GENERATOR

The 555 timer has become a very popular integrated circuit. Its five basic circuits (Fig. 3-14) are two comparators, a bistable flip-flop, a resistive voltage divider, a discharge transistor, and an output stage. The three resistors of the internal divider are of equal value and the threshold comparator is referenced at two-thirds of the supply voltage, while the trigger comparator is referenced at one-third of the supply voltage. The comparators connect to the bistable flip-flop.

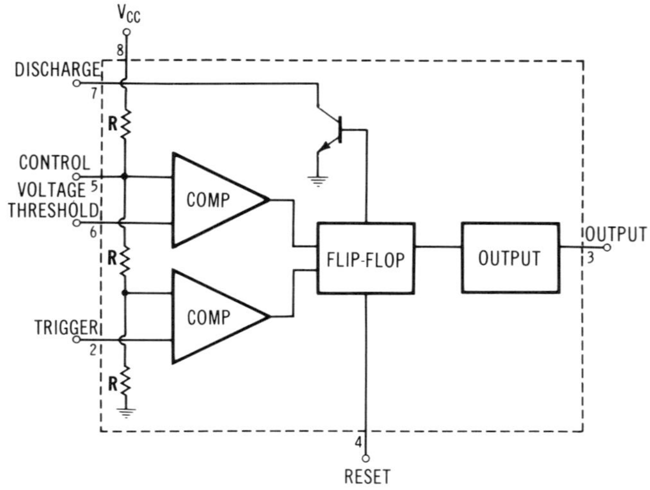

Fig. 3-14. Functional diagram of 555 timer.

When the trigger voltage applied to the trigger comparator drops below one-third of the supply voltage, the comparator changes state and causes the flip-flop to drive the output to its high state. The threshold input monitors the capacitive voltage of an external *RC* timing network. When the capacitor voltage of this external circuit exceeds two-thirds of the supply voltage, the threshold comparator resets the flip-flop. In this case the flip-flop now drives the output to its low state. In this condition the discharge transistor is on and the external timing capacitor will be discharged. When the capacitor is discharged, however, the circuit waits until another trigger pulse arrives to start the timing cycle anew.

When the timer is connected as a pulse generator the threshold and trigger inputs are tied together and are connected to the external capacitor as shown in Fig. 3-15. External resistors are connected to the supply voltage and present a series path along with the capacitor from the supply voltage to common. The discharge circuit is connected to the junction of the two resistors, while the threshold and trigger connections are made between the junction of the second resistor and the capacitor. In operation the application of power to the circuit causes the capacitor to charge toward the supply voltage through resistors R_A and R_B. When the capacitor charge reaches the thresh-

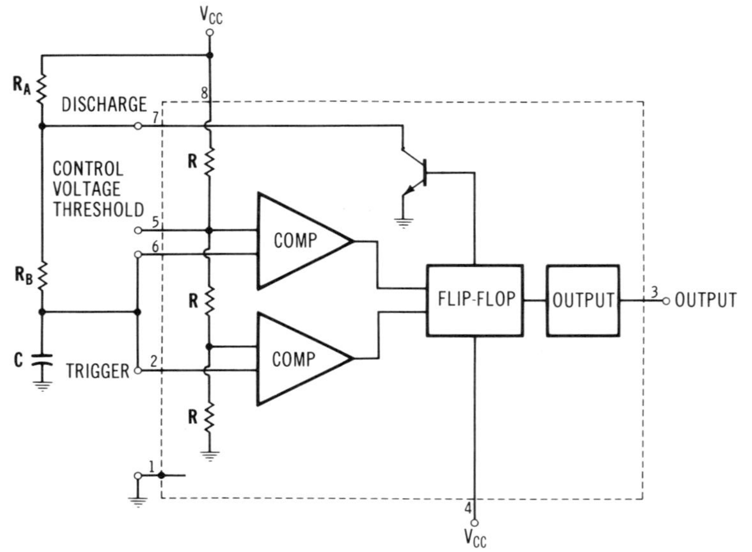

Fig. 3-15. Basic 555 pulse generator.

old level of $\frac{2}{3}V_{CC}$, the output drops low and the voltage on the capacitor discharges through resistor R_B and the discharge transistor of the timer. When the capacitor drops to the level of $\frac{1}{3}$ V_{CC}, the trigger comparator triggers, causing the flip-flop to change over and a new charging cycle to begin. Thus the actual voltage on the capacitor varies between $\frac{1}{3}$ V_{CC} and $\frac{2}{3}$ V_{CC}. The output voltage at pin 3 is high (value of V_{CC}) during the charging interval and low (0 volts) during the discharge period.

The approximate repetition rate or frequency f of the timer is:

$$f = \frac{1.46}{(R_A + 2R_B)C}$$

where R_A, R_B, and C are as shown in Fig. 3-15. The graph of Fig. 3-16 shows the approximate relationship of capacitance, frequency of operation, and the quantity $R_A + 2R_B$. The actual File 3-9 circuit is given in Fig. 3-17. A clean pulse is developed at the output. You may wish to determine the frequency by graph and equation. Use an oscilloscope or frequency counter to determine the actual output pulse repetition rate of your timer.

In operation it is the combined influence of *C1*, R_A, and R_B that determines the frequency of operation. The actual pulse duration is influenced greatly by the value of capacitor *C1* and resistor R_B. If a pulse of a shorter duration but same repetition rate is desired, it can be accomplished by increasing the ohmic value of R_A and decreasing the ohmic value of resistor R_B. You may wish to experiment with different values to evaluate the influence on frequency and pulse duration. The actual equation for duty cycle D is:

$$D = \frac{R_B}{R_A + 2R_B}$$

where R_A and R_B are positioned as shown in Fig. 3-17. This duty cycle can be maintained at the same time a frequency change is made by changing only the value of the capacitor. For example, if the capacitor value is increased by a factor of 10, it will result in a drop in frequency equal to one-tenth of the original value. The duty cycle will not change.

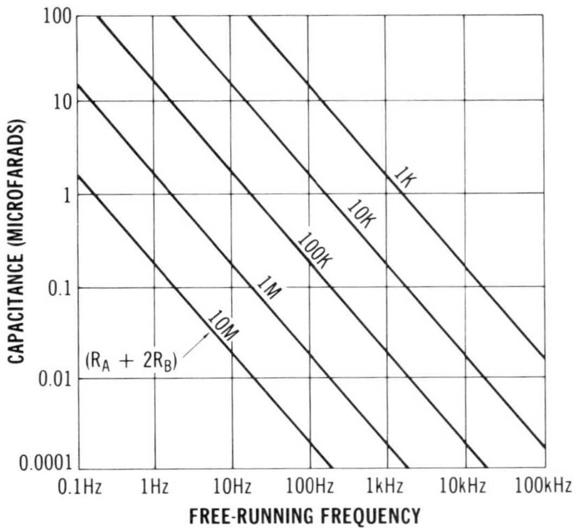

Fig. 3-16. Frequency-capacitance graph for five values of $R_A + 2R_B$.

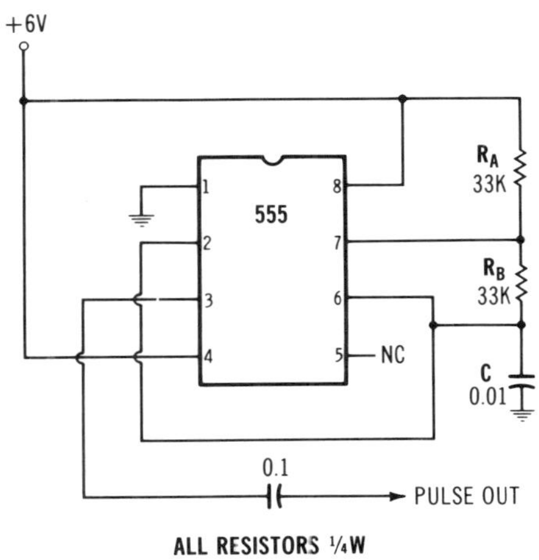

Fig. 3-17. Astable pulse generator.

FILE 3-10

ADJUSTABLE-FREQUENCY SQUARE-WAVE GENERATOR

The circuit of Fig. 3-18 provides a square-wave output regardless of the values for the time-constant resistance and capacitance. Frequency can be adjusted with the potentiometer, *R1*. In this circuit, resistors *R1, R2,* and *R3* in series determine the time-constant resistance, *R*. The time-constant capacitance, *C*, is determined by capacitor *C1*.

During the charge interval of capacitor *C1*, there is a build up toward the supply voltage through the time-constant resistance. Thus the output is high (positive) for an interval of time equal to

$$T_{ch} = 0.693RC$$

where *R* and *C* are the time-constant resistance and capacitance.

The output goes low when the capacitor voltage again reaches the $\frac{2}{3}V_{CC}$ value. However, the discharge path time constant is the same as the charge time constant:

$$T_{dis} = 0.693RC$$

The square-wave frequency becomes

$$f = \frac{1}{1.386RC}$$

The range of frequency operation can be extended by switching in various values of capacitor *C1*.

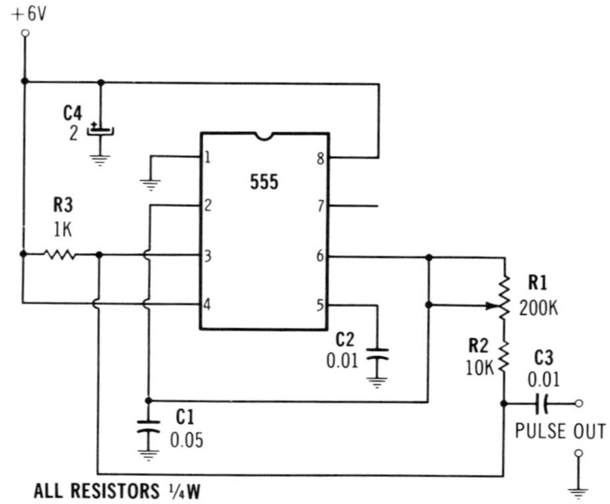

Fig. 3-18. Adjustable-frequency square-wave generator.

555 CONSTANT-FREQUENCY AND VARIABLE-DURATION PULSE GENERATOR

In the circuit of Fig. 3-19 either positive or negative pulses can be generated with adjustable duration. Changing the duration does not change the frequency of operation.

The frequency of operation is determined by the capacitor and resistor values. The potentiometer circuit is such that the net resistance and capacitance values do not change as the potentiometer is varied. The pulse duration, however, does change as the relative charge and discharge times are influenced by the setting of the potentiometer. The frequency of operation can be changed, if desired, by substituting different values for capacitor *C1*.

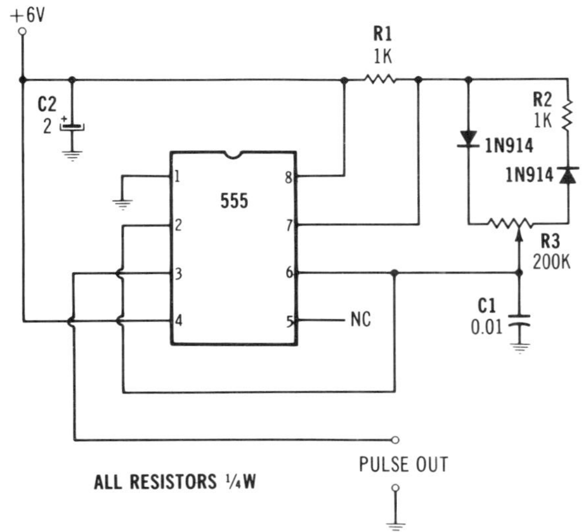

Fig. 3-19. Pulse generator with duration control.

FILE 3-12

100-kHz 555 CRYSTAL OSCILLATOR

The 555 timer IC can be crystal controlled as shown in the circuit of Fig. 3-20. The *R1C1* time constant is set near to the desired frequency of operation. However, the presence of the crystal stabilizes operation at the crystal frequency. If very precise operation is desired, a small variable capacitor can be used in series with the crystal. In checking the author's circuit on WWV at 20 MHz the output was only off frequency by approximately 1.2 kHz. This would be about 6 hertz at the 100-kHz fundamental frequency.

A short-duration negative pulse is obtained at the output. A pulse of longer duration can be obtained by maintaining the same time constant but decreasing the value of *R1* and increasing the value of *C1*. For example, a 1-kilohm resistor for *R1* and a 0.01-μF value for capacitor *C1* will produce a longer-duration negative pulse at the output. The values of *R1* and *C1* have some influence on the frequency of operation of the crystal. There is less shunting effect across the crystal when capacitor *C1* is decreased in value.

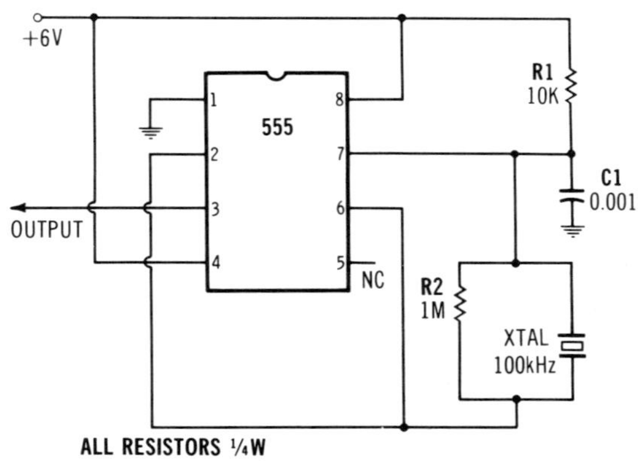

Fig. 3-20. A 100-kHz crystal oscillator.

PULSE POSITION MODULATOR

In a pulse position modulation system the spacing between pulses is made to vary with the modulation (Fig. 3-21). On the trough of the modulating sine wave, for example, the pulses position themselves very near to each other. During the opposite alternation, on the crest, the pulses space themselves widely. In Fig. 3-21 it can be seen that the more negative the modulating sine wave is, the nearer the pulses are to each other. The more positive the modulating wave, the greater is the spacing between pulses.

A basic pulse position modulator using a 555 is shown in Fig. 3-22. This is a basic astable timer circuit with frequency of operation determined by capacitor C and resistors R_A and R_B. The output is removed at pin 3.

A modulating sine wave from an audio generator is applied through a 0.01-μF capacitor to pin 5. In so doing the threshold level to which the capacitor charges varies above the reference dc potential presented by the external time-constant circuit with zero signal voltage at pin 5. As a result the position of the pulses varies with the modulating wave because of the variation in the threshold level.

A change in the amplitude of the sine wave applied to the 555 produces a corresponding change in the separation and nearness of the generated pulses. Thus for a high-amplitude sine wave there will be a greater separation at one extreme and more densely bunched pulses on the opposite alternation as compared to a lower-amplitude modulating wave.

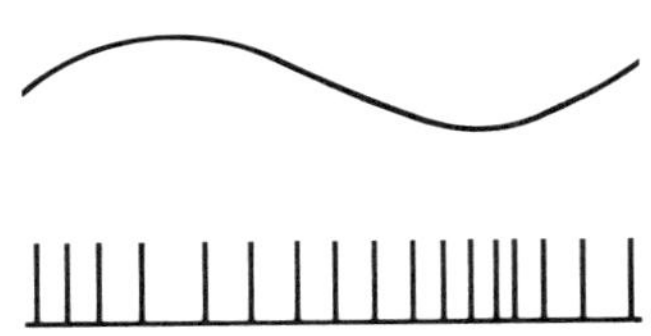

Fig. 3-21. Pulse position modulation.

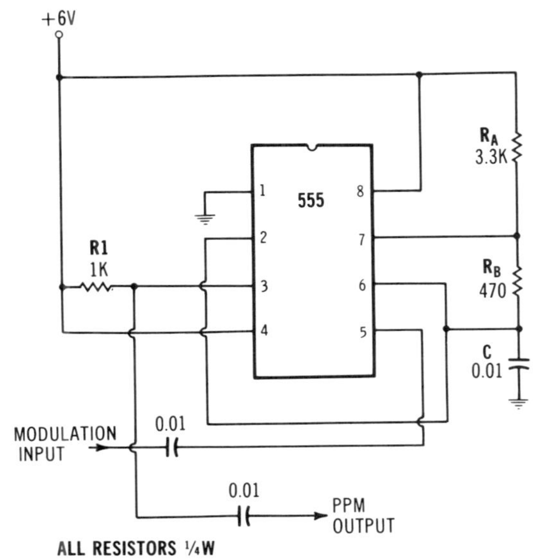

Fig. 3-22. Pulse position modulator.

FILE 3-14

PULSE WIDTH MODULATOR

In a pulse width modulation system the duration of the pulse varies with the modulation (Fig. 3-23). On the crest of the modulating sine wave the pulses are wide. On the opposite alternation they become thinner or shorter in duration. Thus as the modulating wave varies in amplitude the duration of the pulses varies correspondingly.

In a pulse width modulator (Fig. 3-24) the circuit is connected as a monostable instead of a bistable pulse generator. It is necessary to supply a pulse to the clock input, pin 2. The modulating wave, as in the case of the pulse position modulator, is applied to pin 5. The trigger pulse triggers the pulse generator into operation for each arriving pulse. The generated pulse has a duration that depends on the instantaneous amplitude of the modulating wave applied to pin 5. Thus a train of pulses of constant frequency but varying duration is generated. In the author's tests of this circuit the trigger pulse was a 50-microsecond negative pulse derived at pin 3 of the File 3-11 pulse generator.

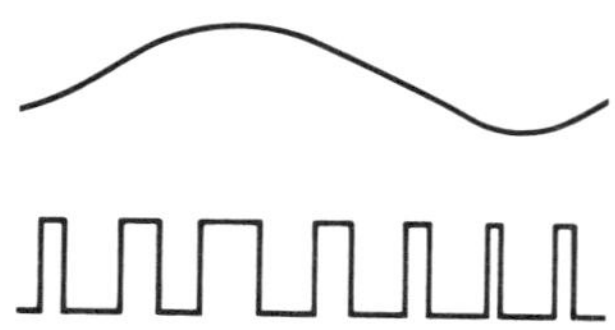

Fig. 3-23. Pulse width modulation.

Note in Fig. 3-23 that the positive crest of the modulating wave produces the longer-duration pulse. The negative trough generates the shortest-duration pulse. If the amplitude of the modulating sine wave is changed, the pulse generated during the positive alternation will be wider and the pulse generated during the negative alternation will be narrower than for a lower-amplitude sine wave.

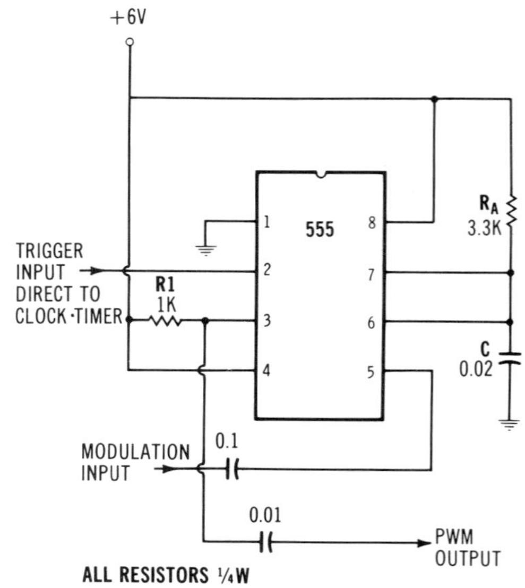

Fig. 3-24. Pulse width modulator.

WAVEFORM GENERATOR: SINE AND SQUARE WAVES

The sine-wave and square-wave generator of File 3-15 uses the Exar XR-2206 monolithic function generator (Fig. 3-25). This waveform generator can be used to form high-quality sine, square, triangle, ramp, and pulse waveforms of high stability. Its possible frequency of operation extends over a range between 0.01 Hz and above 1 MHz.

The waveform generator is comprised of four function blocks, which are: a voltage-controlled oscillator, a multiplier and sine-wave shaper, a unity-gain buffer amplifier, and a set of current switches. In fact, the oscillator frequency can be swept linearly over a 2000-to-1 frequency range with an appropriate external control voltage.

A practical sine-wave and square-wave generator is given in Fig. 3-26. The frequency of operation is controlled externally with a timing capacitor connected between terminals 5 and 6 and appropriate timing resistors connected to terminals 7 and 8. Sine-wave operation is accomplished by connecting a resistor between pins 13 and 14. If an extremely low distortion sine wave is to be generated, an additional potentiometer of 25 kilohms can be used as a symmetry adjustment when connected between pins 15 and 16. Even without this control typical distortion is less than 2.5 percent. The frequency f of operation for the sine-square wave generator is approximately:

$$f = \frac{1}{RC}$$

In the example of Fig. 3-26 the quantity R is the sum of $R3$ plus $R4$, and the capacitance C is equal to the value of $C1$.

A good-quality sine wave was obtained over the entire frequency range using the values indicated for resistors $R1$ and $R2$. If the sine wave is flattened, it may be necessary to adjust the values of resistors $R1$ and $R2$. For the value of $C1$ given the potentiometer varied the frequency over a range of about 200 Hz to more than 100 kHz. The use of three capacitors of values 0.001, 0.01, and 0.1 μF provided overlapping ranges extending from less than 10 Hz to more than 1 mHz. The sine-wave output was approximately 3 volts peak-to-peak. The square-wave output approached 12 volts when a 12-volt supply voltage was used.

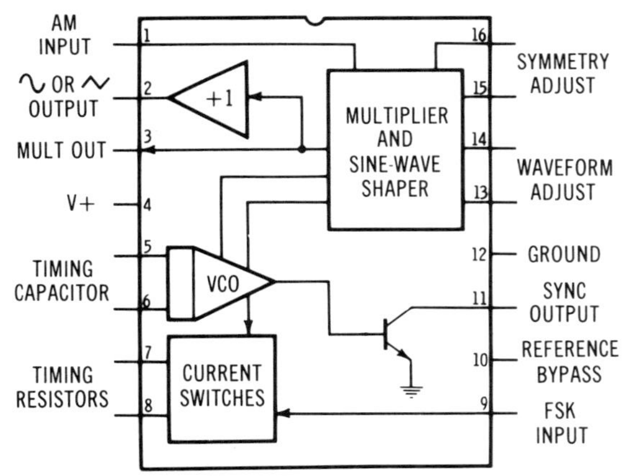

Fig. 3-25. Functional diagram of XR-2206C.

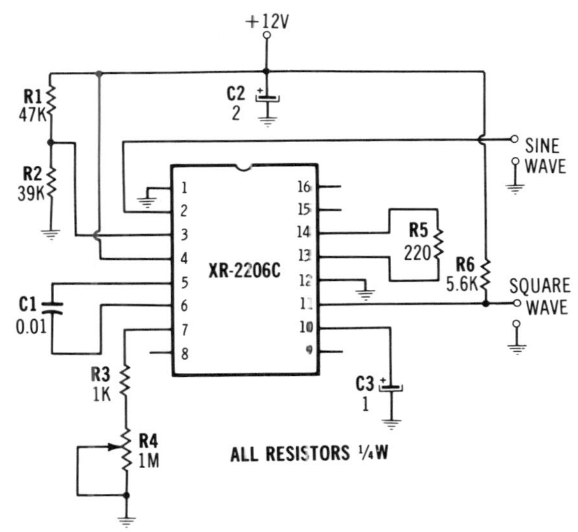

Fig. 3-26. Sine- and square-wave generator.

FILE 3-16

PULSE AND RAMP GENERATOR

The XR-2206 function generator described in Fig. 3-15 can also be connected to generate pulse and ramp (sawtooth) waveforms. This circuit is given in Fig. 3-27. The sawtooth waveform is made available at pin 2, the pulse output, at pin 11. Again the frequency-determining capacitor is connected between pins 5 and 6. The frequency-determining resistors are associated with pins 7 and 8. The frequency f of operation is:

$$f = \frac{2}{(R1 + R2)C1}$$

and the duty cycle D is:

$$D = \frac{R1}{R1 + R2}$$

where *R1*, *R2*, and *C1* are positioned as shown in Fig. 3-27.

The operation of the circuit can be best evaluated by connecting a dual-trace oscilloscope to the two outputs. In the schematic of Fig. 3-27 *R1* has a value of 100 kilohms and *R2* a value of 10 kilohms. These connections result in a positive sawtooth and a negative pulse output.

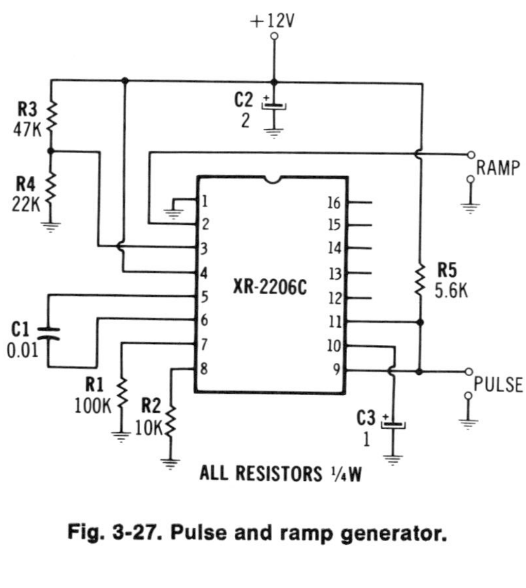

Fig. 3-27. Pulse and ramp generator.

The quality of the sawtooth is determined by resistors *R3* and *R4*. If a straight-line linear sawtooth is not obtained, it may be necessary to adjust these two values.

Interchange the values of resistors *R1* and *R2*. In so doing a negative sawtooth and a positive pulse output is obtained. Make resistors *R1* and *R2* of equal 100-kilohm value. These connections produce a triangle-wave output and a square-wave output.

To obtain an extremely short duration output pulse you require a greater ohmic separation between the values for resistors *R1* and *R2*. Substitute a 100-kilohm value for *R1* and 1 kilohm for *R2*. Note the very short duration negative pulse that is obtained.

A control can be used to change the frequency of operation as shown in Fig. 3-28. A 1-megohm resistor is connected in series with the 100-kilohm resistor used for *R1*. Resistor *R2* has a value of 10 kilohms. Note that the frequency of operation of the function generator changes as the 1-megohm resistor is varied through its range.

Different frequency ranges can be established by changing the value of capacitor *C1*. Use a value of 0.001 μF. In so doing a much higher frequency range can be tuned by the 1-megohm potentiometer. For low-frequency operation, substitute a capacitor value of 0.1 μF. Overlapping frequency ranges are obtained, but with the three values of capacitance you can obtain outputs from 20 Hz to 20 kHz.

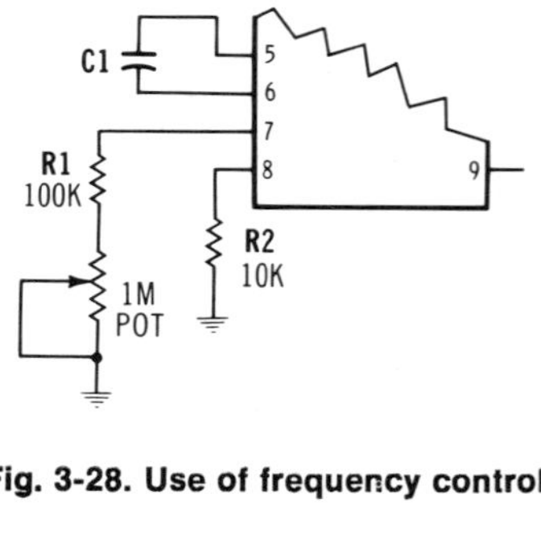

Fig. 3-28. Use of frequency control.

CODE-PRACTICE OSCILLATOR AND TONE GENERATOR

A 555 timer can be connected as a simple audible-tone generator and code-practice oscillator. It has a pleasant tone and keys well. The basic circuit is given in Fig. 3-29.

The frequency of operation is determined by resistors *R1, R2,* and *R3* plus capacitor *C1*. Potentiometer *R1* varies the tone frequency over a range of frequencies that are most suitable for nonirritating code practice. Keying is done in the ground return associated with pin 1.

The volume level can be regulated with resistor *R4*. The highest output is obtained with resistor *R4* shorted out of the circuit. A quiet, low-level tone is obtained by using a resistor value of approximately 220 ohms.

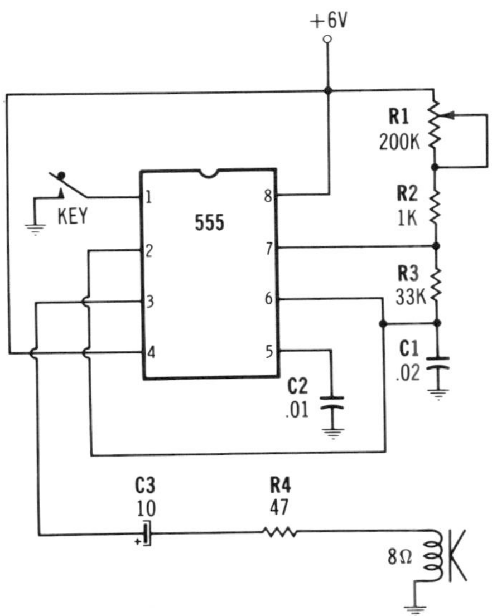

Fig. 3-29. Code-practice oscillator.

FILE 3-18

555 SQUARE-WAVE GENERATOR

The time-constant circuit associated with a 555 astable oscillator can be rearranged to obtain a square-wave output as shown in Fig. 3-30. Note the position of resistor *R2* between pin 7 and the junction of resistor *R1* and capacitor *C1*. If this resistor has an ohmic value that approximates one-half the ohmic value of resistor *R1*, the charge and discharge time constants become nearly equal, setting up a square-wave output (duty cycle of 50 percent). It is important that the ohmic value of resistor *R2* is not greater than one-half of the ohmic value of resistor *R1* if oscillations are to be produced. Making resistor *R1* substantially greater in ohmic value than resistor *R2* results in a negative output pulse rather than a square wave.

The frequency of operation can be changed by using different values for resistors *R1* and *R2*, always keeping resistor *R1* approximately double the ohmic value of resistor *R2*. The frequency of operation can also be changed by using a different value for capacitor *C1*.

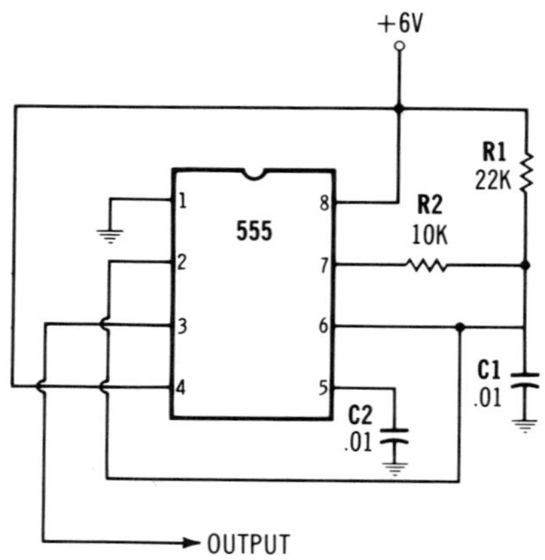

Fig. 3-30. Square-wave generator.

LED FLASHER AND OSCILLATOR

The LM3909 can activate light-emitting diodes. Only two external components are needed (Fig. 3-31) : the LED and a charging capacitor. Timing resistors and the LED current-limiting resistor are internal. The integrated circuit operates at low voltage (1 to 5 volts) and with a very low power drain. Operating continuously, the circuit will operate for more than one year using a single size C flashlight cell. The current drain may average less than 0.5 milliampere.

The timing capacitor charges and delivers 2 volts to the LED for flashing while operating with a supply voltage of 1.5 volts or less. A 300-μF capacitor will result in a flashing rate of about one flash per second. The capacitor need only have a 3-volt rating. A larger capacitor will slow down the flashing rate; a smaller capacitor will increase that rate. A 1000-μF capacitor resulted in approximately 20 flashes per minute.

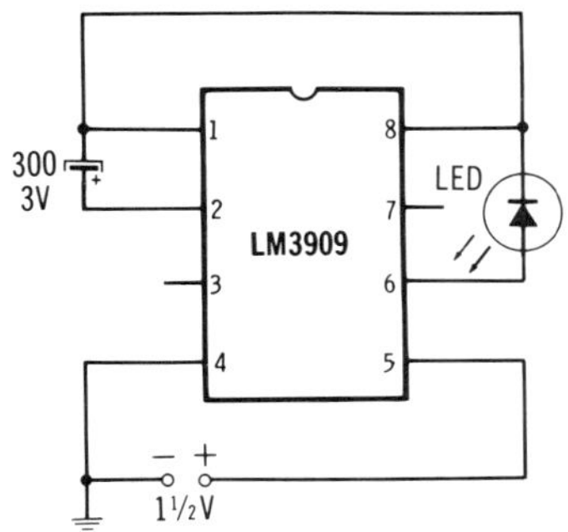

Fig. 3-31. LED flasher.

A 100-kHz crystal substituted for the capacitor will function as a square-wave oscillator with output available at pin 2. The LED must be removed from the circuit.

A low-voltage and low-power-drain square-wave generator is given in Fig. 3-32. The components selected establish a frequency in the 1000- to 1500-hertz range, depending on parts tolerances. The output is at pin 2. The frequency of operation is determined by capacitor *C1* and resistor *R1*. Decreasing the value of capacitor *C1* will increase the frequency of operation.

Square-wave or pulse output can be determined by the ohmic value of resistor *R3*. If this resistor is shorted out of the circuit a very narrow negative pulse is produced instead of a square wave.

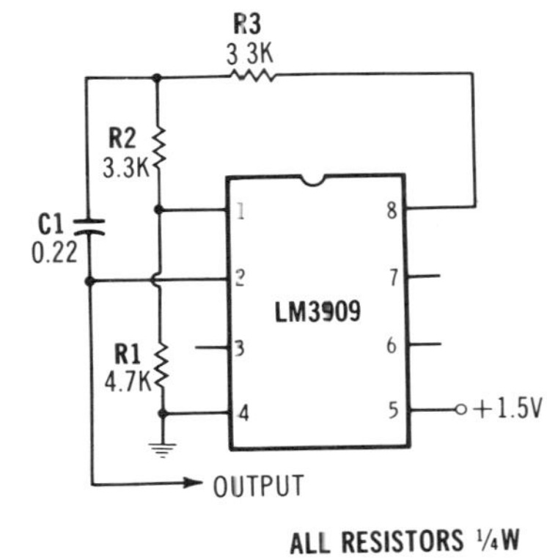

Fig. 3-32. Low-voltage and low-power square-wave generator.

FILE 3-20

LOW-FREQUENCY GENERATOR

The Exar XR-2240 timer/counter operates at very low frequencies as a generator, counter, or time-delay circuit. Its timing operations can be controlled from microseconds to days using programmable delays that extend between $1RC$ to $255RC$. This timer integrated circuit can also be used as a frequency synthesizer.

The XR-2240 (Fig. 3-33) consists of a time base generator (similar to a 555) and a binary counter that can provide selected counts of 1, 2, 4, 8, 16, 32, 64, and 128. A control system and suitable external circuits permit operation in monostable or bistable modes. Operation can be initiated by a positive trigger pulse and stopped by a reset pulse.

The XR-2240 connected as a low-frequency generator is shown in Fig. 3-34. The timer frequency is set by resistor *R1* and capacitor *C1*. For the values shown the operating frequency f would be:

$$f = \frac{1}{R1C1}$$

$$= \frac{1}{1 \times 10^6 \times 0.1 \times 10^{-6}} = 10 \text{ Hz}$$

The operating frequency can be checked with a counter or oscilloscope. Precise operation depends on the use of close-tolerance components for resistor *R1* and capacitor *C1*. These close tolerances are not necessary for routine checking of the operation of the circuit. The recommended range of timing components suggested for resistor *R1* and capacitor *C1* are:

Resistor *R1*:	1000 ohms to 10 megohms
Capacitor *C1*:	0.007 μF to 1000 μF

The binary counter chain that follows the time base oscillator provides a total count of 128. A very short duration negative pulse that can be seen at pin 14 is supplied to the binary counter. When the XR-2240 is operated as a free-running oscillator the output at pin 1 is one-half the frequency of the time base oscillator output. Consequently, the output of each section is halved as well. The output at pin 8 is a square wave that is 1/256th of the time base oscillator frequency.

In the circuit arrangement of Fig. 3-34 individual count signals are connected to the output by closing the appropriate switch. For example, a division of 32 is obtained by closing the switch associated with pin 5.

When operating at such a low frequency the output square waves will have periods measured in seconds (output frequencies of less than 1 hertz). Circuit operation at these low frequencies is best observed on an oscilloscope with the vertical amplifier set for dc operation. With the pin 8 output active you will notice the shift of the baseline of the oscilloscope every so many seconds. In the author's setup a full cycle of operation of the output at pin 8 required approximately 30 seconds or about 2 cycles per minute. Substituting larger val-

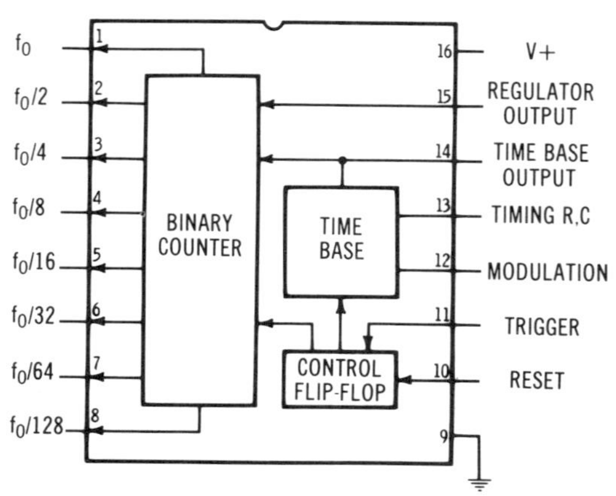

Fig. 3-33. Functional diagram of XR-2240 timer/counter.

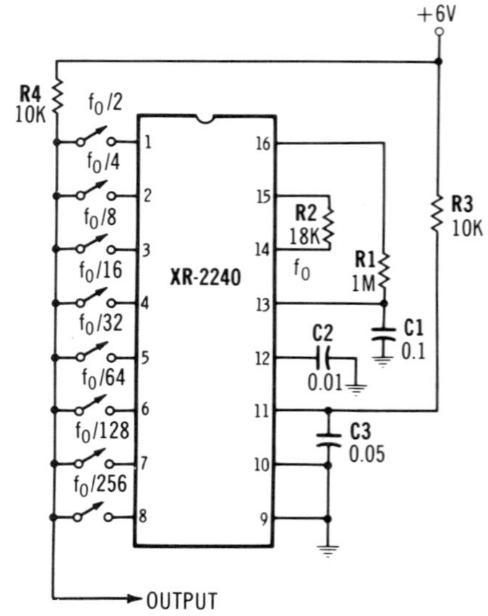

Fig. 3-34. Low-frequency generator.

ues of capacitor for *C1* can extend the period up into minutes, fractional parts of an hour, and hours.

The XR-2240 can be made to operate at higher frequencies by changing the time constant of the time base generator. Substitute a value of 10 kilohms for resistor *R1* and 0.01 μF for *C1*. The operation will now be in the audio-frequency range and it will be possible to use an oscilloscope to observe each of the outputs, at pins 1 through 8. Use an oscilloscope to observe each output individually. Note how the frequency halves as you proceed from the outputs at pins 1 through 8. Each of these pins provides a square-wave output. The final output will be 1/256th of the time base oscillator frequency.

Many unusual pulse output waveforms can be obtained by switching more than one pin to the output. Note that with pins 1 and 3 both connected to the output the resultant signal is short-duration positive pulse pairs. If pins 1 and 4 supply output, the pulses appear in groups of four. When operating in this manner the actual pulse pattern repeats itself at a rate equal to the period of the lowest frequency (highest counter bit connected to the output). The minimum pulse width obtained is set by the pulse width for the highest frequency output (lowest counter bit connected to the output).

Note the waveform displayed when connecting more than two pins to the output. For example, with pins 1, 3, and 5 connected there will again be pulse pairs. The pairs, however, will either be spaced narrowly or widely from each other because of the output contributed at pin 5. Recall that when using pins 1 and 3 alone there was equidistant spacing between the pulse pairs. You can observe an extensive variety of pulse output waveforms by experimenting with various output combinations.

FILE 3-21

LONG TIME DELAY CIRCUIT

The XR-2240 timer described in File 3-20 can be used as a short- or long-duration time delay circuit. In this delay function the timer can be operated in monostable or bistable manner. When switch *S1* of Fig. 3-35 is closed, monostable operation results; switch *S1* open provides bistable activity.

The fundamental delay time is set by the *R1C1* time constant associated with the time base oscillator. The time constant for the values given in Fig. 3-35 is:

$$T_o = R1 \times C1$$
$$= 1 \times 10^6 \times 10 \times 10^{-6} = 10 \text{ seconds}$$

When the switch associated with pin 1 of the counter circuit is closed the delay time T_T at the output will be 10 seconds. When the switch associated with pin 2 is closed and the remaining switches are open the delay time T_T at the output will be 20 seconds or $2T_o$. Other times are indicated in Fig. 3-35.

The individual switches associated with the counter outputs need not be used alone. They can be used together to establish other delay values. For example, assume that a 30-second delay is desired. This can be obtained by closing the switches associated with pins 1 and 2. Then the total time is $(1 + 2)T_o$ or $3T_o$.

To observe the operation of the circuit a voltmeter or oscilloscope connected for dc operation is connected to the output. Prior to the connection of a positive trigger pulse to pin 11 the output will be *high,* approaching the supply voltage. As soon as the trigger is applied the output will go *low* (0 volts) and hold this value for the amount of the delay time.

If switch *S1* is in closed position the output will now maintain a high output until the next trigger pulse is applied to pin 11. This provides monostable operation of the delay circuit. If at any time during the delay interval a positive reset pulse is applied to pin 10, the output will go back to high.

Astable operation is obtained by opening switch S1. Again the activities are initiated by applying a positive trigger to pin 11. After the specific delay time the output will again go high. In bistable operation, however, the output will again go back low, even without the application of a trigger pulse to pin 11. In other words, in the bistable operating mode, once the activity is initiated by an applied trigger pulse, the circuit will recycle itself and continue operation at the rate determined by the switched-in delay time. This bistable mode of operation can be stopped by applying a positive reset pulse to pin 10.

To observe the operation of the circuit an applied trigger pulse can be initiated by momentarily closing switch *S2* to the supply voltage. Remember that only a momentary contact must be made; do not hold the switch closed. Use a watch with a second hand to measure the timing intervals. Close the switch associated with pin 1. Momentarily close switch *S2* to the supply voltage. The output will immediately go to zero. Use the watch to determine how long the output will remain at zero before it swings back to the supply voltage value. Depending on parts tolerances associated with the timer this should be approximately 10 seconds.

Repeat the preceding with the switches associated with pins 1 and 2 both closed. Now the out-

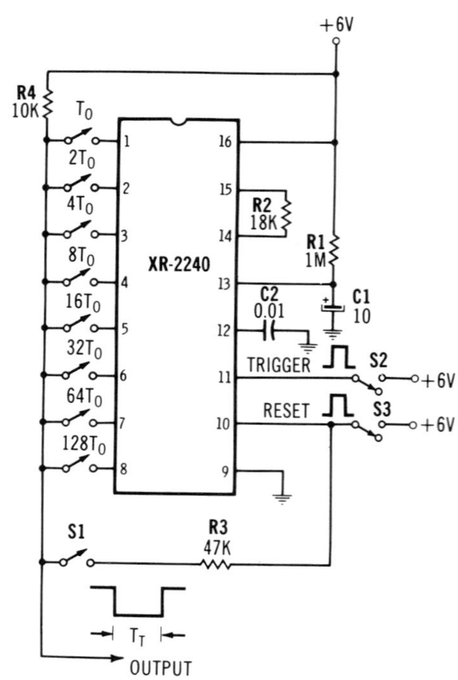

Fig. 3-35. Time-delay circuit.

put should go to zero for a time interval of 30 seconds.

Initiate a 30-second delay interval by closing switch *S2* momentarily. After you have started this delay, momentarily close switch *S3* to +6 volts. This applies a reset pulse and immediately returns the output to the high value. This step demonstrates the activity of a reset pulse in returning the timer to a start position.

What counter switches need to be closed to obtain a 1-minute delay time? This amount of delay can be accomplished by closing the switches associated with pins 2 and 3. Subsequent to the application of the trigger by momentarily closing switch *S2* to the supply voltage, the output will remain low for 60 seconds (1 minute).

How would you establish a delay time of 10 minutes (600 seconds)? This can be accomplished by closing the switches associated with pins 3, 4, 5, and 6, producing accumulated delays of 40, 80, 160, and 320 seconds. Check out the circuit operation for this amount of delay.

Other delay times can be established by using the appropriate switches associated with pins 1 through 8. Furthermore, other delay combinations can be made by changing the basic period of the time base oscillator. For example, a basic T_0 period of 100 seconds can be obtained by using a 1-megohm resistor for *R1* and 100-μF capacitor for *C1*.

Remember that a one-shot or monostable delay time of a specific value can be set up by closing switch *S1*. If the delay interval is to be repeated time and time again in astable fashion, switch *S1* should be in its open position. In this bistable mode of operation the output will remain high (after each delay period) by a time interval corresponding to the time set by the lowest-count switch closed. For example, if the lowest-count switch closed is the one at pin 2, the high time interval will be $2T_0$. After this time the cycle will again switch to the low time-delay duration, starting a new cycle of activity.

FILE 3-22

FREQUENCY SYNTHESIZER

The XR-2240 timer/programmer covered in Files 3-20 and 3-21 can also be used as a frequency synthesizer (Fig. 3-36). The integrated circuit operates in a free-running manner and the output frequency depends on the time constant of the time base oscillator and the single or combination of counter switches closed. A series of positive pulses are generated with the pulse duration, regardless of frequency, set by the time constant of the time base generator. For the component values selected in Fig. 3-36 the pulse duration is:

$$T_o = R1C1$$
$$= 10{,}000 \times 0.01 \times 10^{-6} = 0.0001 \text{ second}$$

This corresponds to a pulse duration of 100 microseconds. The basic frequency of the time base oscillator is the reciprocal of the time constant, or:

$$f = \frac{1}{0.0001} = 10{,}000 \text{ hertz}$$

is the basic frequency.

The period of the pulse output depends on which switch or group of switches is closed to the counter outputs, pins 1 through 8. The equation for the pulse period T_o is:

$$T_p = (N + 1)T_o$$

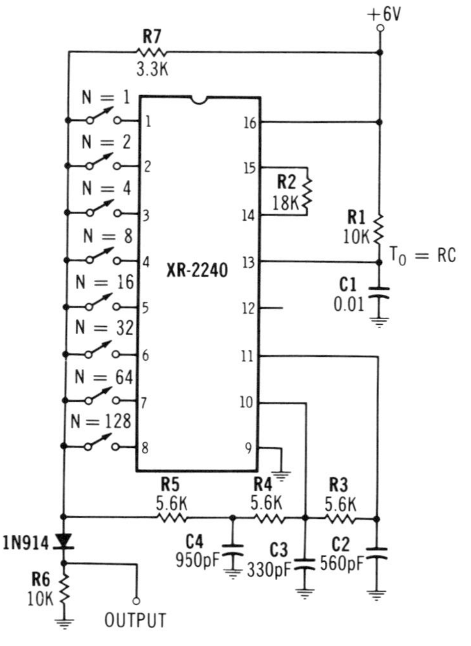

Fig. 3-36. XR-2240 frequency synthesizer.

and the frequency or pulse repetition rate is:

$$f_{out} = \frac{1}{T_p} = \frac{1}{(N+1)T_o}$$

where N is the number specified by the pin with its switch closed (Fig. 3-36).

Several examples demonstrate the output pulse period and frequency as a function of the switch closings. When the switch associated with pin 1 is closed, the period of the output pulse is:

$$T_p = (N + 1)T_o$$
$$= (1 + 1) \times 0.001 = 0.0002 \text{ second}$$

and

$$f_{out} = \frac{1}{T_p} = \frac{1}{0.0002} = 5000 \text{ hertz}$$

is the pulse output frequency.

What is the pulse output frequency when only the switch associated with the pin 3 output is closed? Since $N = 4$ for pin 3,

$$f_{out} = \frac{1}{(4 + 1) \times 0.0001} = 2000 \text{ hertz}$$

How would you obtain an output frequency of 2500 hertz? This is accomplished by closing both switches associated with pins 1 and 2. As a result:

$$f_{out} = \frac{1}{(1 + 2 + 1) \times 0.0001} = 2500 \text{ hertz}$$

An output frequency of 1000 hertz can be obtained by closing the switches for pins 1 and 4.

$$f_{out} = \frac{1}{(1 + 8 + 1) \times 0.0001} = 1000 \text{ hertz}$$

An output frequency of 100 hertz is obtained by closing the switches for pins 1, 2, 6, and 7.

In each case the duration of the output pulse is 0.0001 second but the spacing between pulses varies in accordance with the desired output pulse repetition rate (frequency). The total number of pulse frequencies feasible is 256, depending on the counter switch combinations that are closed. Of course, other pulse frequencies are feasible, too, by changing the time constant of the time base oscillator. This time constant can even be made variable by inserting a potentiometer in series with resistor *R1*.

LONG-RANGE TIMER

The Exar XR-2242 long-range timer is capable of producing ultralong time delays from microseconds to days. The XR-2242 (Fig. 3-37) consists of a time base generator and an 8-bit binary counter that can provide a total count of 128. A control system and suitable external circuits permit operation in monostable or bistable modes. Operation can be initiated by a positive trigger pulse and stopped with a positive reset pulse.

A basic time base interval is set by the resistor-capacitor combination associated with the time base oscillator. The total time delay of the timer is this quantity multiplied by the total count of 128. The output at pin 3 is a negative timing pulse of 128*RC*. After this interval the circuit resets itself to complete the timing cycle. A secondary output is also available at pin 2. This is a square-wave output with a period of 2*RC*.

A complete timing circuit with an associated LED indicator is shown in Fig. 3-38. The time base oscillator frequency is set by resistor *R1* and capacitor *C1*. For the values shown the time constant is:

$$T_o = R1C1$$
$$= 0.47 \times 10^6 \times 0.1 \times 10^{-6}$$
$$= 0.047 \text{ second}$$

The recommended range of timing components suggested for resistor *R1* and capacitor *C1* is:

Resistor *R1*: 1000 ohms to 10 megohms
Capacitor *C1*: 0.007 μF to 1000 μF

The pulse generated by the timing pulse is applied to the counter which provides a total count of 128 to 1. Consequently, for the values given in Fig. 3-38 the total count T_T would be:

$$T_T = 128T_o = 6.016 \text{ seconds}$$

This output is applied to a transistor driver and LED output circuit. The output is high prior to the initiation of the delay time. Consequently the LED is lit. During the delay time the LED is not lit.

Switch *S1* selects either monostable or bistable operation of the timer. When the switch is open the timer operates in bistable fashion. The output is initiated by applying a positive pulse to pin 6 or connecting switch *S2* to the supply voltage momentarily. When this happens the LED light goes out for a duration of time equal to 128*RC*. If at any time during the delay interval it is necessary to reset the timer it can be done by applying a positive reset pulse to pin 5 or momentarily connecting pin 5 to the plus supply voltage. This will also stop bistable operation of the circuit. Activity can be started again by closing switch *S2* momentarily to the supply voltage. Operation then will be initiated again and bistable operation will continue indefinitely until the reset circuit is again activated.

Monostable operation is obtained by closing switch *S1*. Momentarily close switch *S2* to the

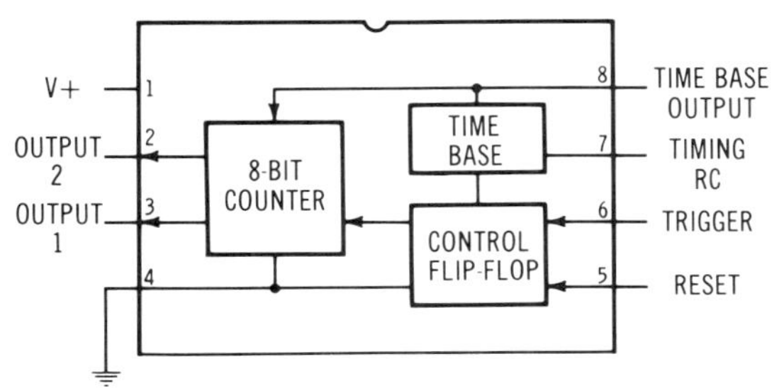

Fig. 3-37. Functional plan of XR-2242 timer.

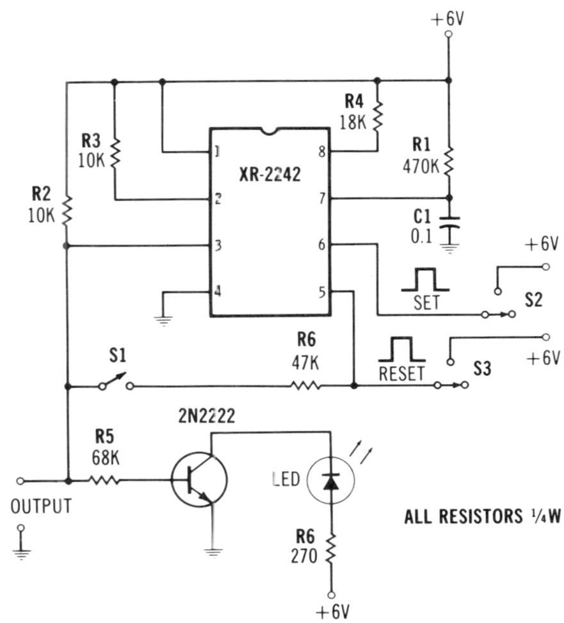

Fig. 3-38. Long-range timer with LED indicator.

supply voltage to initiate circuit activity. The LED will turn off and will remain off for the extent of the delay time. After this time the LED will again light and will remain so until *set* activity is initiated again at pin 6.

Set up the circuit of Fig. 3-38 and measure the output delay time with a stopwatch. It should be approximately 6 seconds. Longer time delay periods can be initiated by increasing the value of the timing capacitor. Substitute a 100-μF capacitor for *C1*. What will the new delay time be?

$$T_T = 0.47 \times 10^6 \times 100 \times 10^{-6} \times 128$$
$$= 6016 \text{ seconds}$$

This corresponds to 100.3 minutes, or approximately 1 hour and 40 minutes. Set up the timer circuit and initiate *set* operation and determine the length of time that the LED remains unlit. The use of a 1000-μF capacitor will establish a delay time of 16⅔ hours.

The timer can be operated at higher frequencies in the audio range by decreasing the time constant associated with the time base oscillator. A 0.01-μF capacitor for *C1* and a 10-kilohm resistor for *R1* would set up a T_o of 0.0001 second. This would correspond to a frequency of 10,000 hertz. An oscilloscope can now be used to observe the waveforms and delay intervals at pins 2 and 3.

REMOTELY CONTROLLED ELECTRONIC ATTENUATOR

The Motorola MC3340P is an integrated circuit that will permit an adjustable attenuator to be remotely controlled. This can be accomplished by a dc control voltage or an external variable resistor. The attenuator can be positioned in the signal path close to input and output connections. A dc control voltage applied to pin 2 can then be used to vary signal amplification such as might be necessary for a compression or expansion amplifier system. In the second application a remotely located potentiometer can be used to vary the amount of signal attenuation from a distant location.

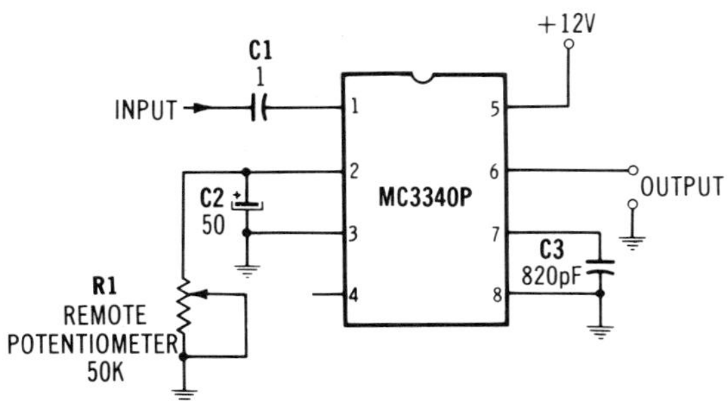

Fig. 3-39. Remotely controlled electronic attenuator.

These control functions can be made without influencing the operating characteristics of the attenuator with the exception of attenuation or gain. In fact, the magnitude of the output can be varied above and below the level of the input signal. The maximum voltage gain of the device is typically +13 dB. The attenuation limit of the device is typically −90 dB. The maximum input voltage is 0.5 volt rms.

A remote volume control circuit is given in Fig. 3-39. The input signal is applied to pin 1, the output removed from pin 7. The remotely located potentiometer connects to pin 2. Apply a low-level, 1000-hertz sine wave to the input and note the influence of the potentiometer setting on the output voltage.

The high-frequency roll-off is set by capacitor *C3*. The value given will provide good audio-frequency response well above 10,000 hertz. If high-frequency response is to be limited to the voice-frequency range, a value of 0.01 μF or higher can be used, depending on the voice-frequency roll-off desired. The electronic attenuator will operate with a supply voltage between 9 and 20 volts; 12 volts is typical.

FILE 3-25

DOT/BAR DISPLAY DRIVER

The National Semiconductor LM3914 is able to sense analog input voltages applied to pin 5 (Fig. 3-40) in such a manner that ten light-emitting diodes (LEDs) provide a linear display of the input voltage level, be it constant or changing. The circuit can be adjusted to display a bar graph or a moving dot pattern.

Internal to the integrated circuit is an accurate ten-step voltage divider and an adjustable reference that can be set between 1.2 and 12 volts. Associated with the divider chain are ten individual comparators that respond to the fixed or changing voltage level on the dividers and turn on the corresponding LEDs.

Supply voltages are connected to pins 2 and 3. Pins 4 and 6 represent connections to the low and high ends of the divider chain. Pin 7 is the reference 1.2-volt output of the device. An appropriate circuit in connection with pins 7 and 8 can be used to set the actual reference level in the operation of the display driver. When pin 8 is not used and the reference is set by the output of pin 7, the reference voltage is 1.2 volts. The application of this voltage to pin 5 would result in the illumination of all LEDs for a bar graph display and the illumination of LED 10 for a dot display. An applied voltage of 0.12 volt to pin 5 would cause illumination of LED 1 only. The application of 0.5 volt to pin 5 would cause the illumination of LED 4 (0.12×4) for a dot display and the illumination of LEDs 1, 2, 3 and 4 for a bar graph display.

A demonstration display circuit is given in Fig. 3-41. Note how the LEDs are arranged in the circuit by their anodes being connected to the supply voltage. The dot/bar operation is controlled with switch *S1*. Pin 9 is connected to pin 11 for a dot display. Pin 9 is connected to the supply voltage for a bar graph display.

The reference voltage is set by resistors *R2* and *R3* connected to pins 7 and 8. The reference voltage is determined by the following equation:

$$\text{reference voltage} = 1.25\left(1 + \frac{R3}{R2}\right)$$

where *R3* and *R2* are as shown in Fig. 3-40. In the circuit of Fig. 3-41 the reference voltage is approximately 5.5 volts. Therefore, if 5.5 volts is applied to pin 5, LED 10 would turn on with the circuit arrangement shown. Were switch *S1* set to the bar position, all LEDs would be illuminated.

In the demonstration circuit the voltage for application to pin 5 is derived from the reference voltage source. Note that the one side of the potentiometer *R1* connects to pin 7 and the other side to ground. Thus as the arm of the potentiometer is varied, the voltage applied to pin 5 changes between 0 and about 5.5 volts. If you

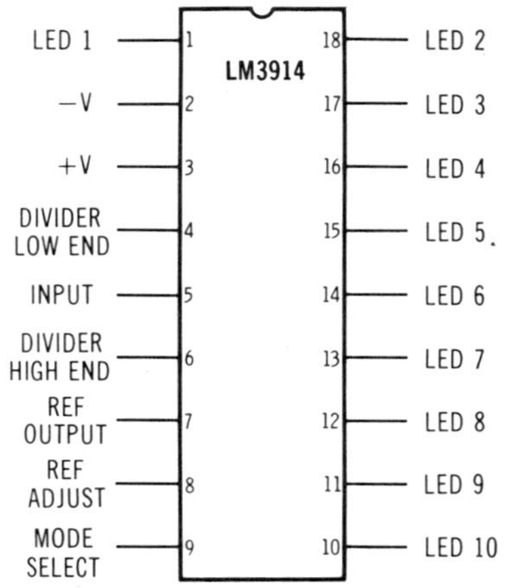

Fig. 3-40. Pin-out diagram for dot/bar display driver.

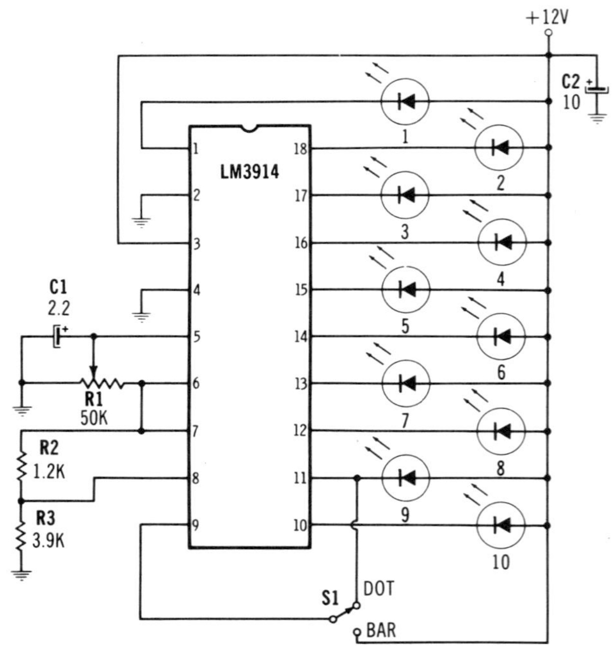

Fig. 3-41. Dot/bar display circuit.

vary the potentiometer from 0 volts to 5.5 volts the LEDs will light progressively only one at a time between LED 1 and LED 10 with switch *S1* set to the dot position. Were *S1* set to the bar position, the same progression would occur except that all the previous LEDs illuminated would remain lit.

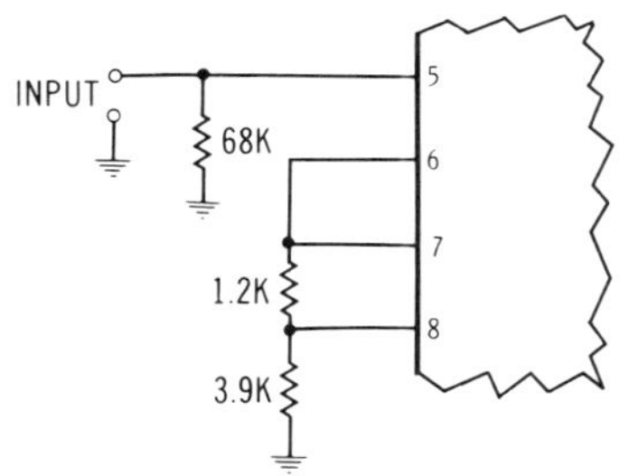

Fig. 3-42. Connections for measuring an external voltage.

To measure an external voltage connect pins 5, 6, 7, and 8 as shown in Fig. 3-42. Pin 5 is no longer associated with the reference voltage source and an external voltage can now be applied for measurement.

Connect a small 1.5-volt battery between pin 5 and ground (positive side to pin 5). Note that LED 2 will glow, indicating that the voltage is greater than 1.1 (2 × 0.55) volts but lower than 1.65 volts (3 × 0.55). Switch over to the bar display and note that both LED 1 and LED 2 are lit.

Connect a 6-volt battery between pin 5 and ground. This will illuminate LED 10 for the dot display and will illuminate all LEDs for the bar display.

NOTES

NOTES

NOTES

NOTES